Encyclopaedia of
Sustainable Water Resource Management

Encyclopaedia of

Sustainable Water Resource Management

Volume 4

Editor:

R.N. Sahay

AGROTECH PRESS

JAIPUR ▪ NEW DELHI

AGROTECH PRESS

Regd. Off. "Bony Residency", Gate No. 2
Opp. Tilak Public School, Vishweswariya Nagar,
Gopalpura Road, Jaipur - 302018, India
Ph: 0141 - 2761280, Telefax: 2761381
E-mail : agrotechpress@gmail.com

Delhi Office: 102, 1st Floor, Satyam House,
4327/3, Ansari Road, Darya Ganj,
New Delhi - 110002, India
Ph: 011-45652440
visit us: www.agrotechpress.com

Distributed By
KEYA INTERNATIONAL PRESS LLP
E-mail: keyainternationalpressllp@gmail.com

Published in India, 2025

ISBN : 9789384568269 (5 Vols. Set)

Typeset by Thomson Press

Printed and bound in India by Thomson Press Ltd.,
New Delhi

Contents

Preface

Water is an essential resource for all life on the planet. Of the water resources on Earth only three per cent of it is not salty and two-thirds of the freshwater is locked up in ice caps and glaciers. Of the remaining one per cent, a fifth is in remote, inaccessible areas and much seasonal rainfall in monsoonal deluges and floods cannot easily be used. At present only about 0.08 per cent of all the world's fresh water is exploited by mankind in ever increasing demand for sanitation, drinking, manufacturing, leisure and agriculture.

Much effort in water management is directed at optimising the use of water and in minimising the environmental impact of water use on the natural environment. Successful management of any resources requires accurate knowledge of the resource available, the uses to which it may be put, the competing demands for the resource, measures to and processes to evaluate the significance and worth of competing demands and mechanisms to translate policy decisions into actions on the ground. For water as a resource this is particularly difficult since sources of water can cross many national boundaries and the uses of water include many that are difficult to assign financial value to and may also be difficult to manage in conventional terms. Examples include rare species or ecosystems or the very long term value of ancient ground water reserves. The field of water resources management will have to continue to adapt to the current and future issues facing the allocation of water. With the growing uncertainties of global climate change and the long term impacts of management actions, the decision-making will be even more difficult. It is likely that ongoing climate change will lead to situations that have not been encountered. As a result new management strategies will have to be implemented in order to avoid setbacks in the allocation of water resources. In water resources management, available water resources are allocated between different needs and demands. Where water is scarce this presents a considerable challenge. Usually water is divided between different users - agriculture, industry, hydro-power generation, water supply, mining, transport requirements, tourism etc. Often the water which is required for aquatic ecosystems is rather grudgingly allocated as an after-thought and often only after those making the allocations are forced to do so through regulatory requirements. In recent years, however, riverine ecosystems have been increasingly identified as the 'source of the resource' which perform a number of tasks which are of great value and which are essential for the sustainability of the system. Integrated water resources management is the "maintenance of ecosystem integrity, according to a management principle of preserving aquatic ecosystems, including living resources, and of effectively protecting them from any form of degradation on a drainage basin basis". It also recommends the adoption of an integrated approach to environmentally sustainable management of water resources, including the protection of aquatic ecosystems and freshwater living resources, and the integration of water quality elements into water resources management. However, current integrated water resources management practices often consider ecosystems primarily as water users, with little attention given to their vital role as providers and regulators of water resources. Little consideration has been given to the other services and goods that ecosystems provide, such as flood regulation, Biodiversity conservation, fish and firewood. It is of fundamental importance to the long term availability and sustainable management of water resources that the maintenance of ecosystems and the strengthening of their role as providers of services and goods be recognised. Similarly, in spite of high-level of commitments to action made at both the International Conference on Water and the Environment and the Earth Summit, water quality has invariably been subordinated to water quantity and sanitation has been neglected. The result of that neglect is an emerging crisis of water quality, damaging public health and restricting economic development.

This book presents views and opinions of experts regarding conservation and effective utilization of water resources that are fast becoming scarce in their availability. The gravity of the problem, the realities and consequences on the present and future generations is a well-recognized fact which calls for rational utilization of water resources. The book will be of interest to a wide readership including researchers, scientists, academicians, engineers, planners, policy-makers, eco-technologists and environmentalists.

—Editor

1

The Clean Water Act: Strategies

The Clean Water Act (CWA) has proven to be a remarkably effective and adaptive law over its forty-year history. It is widely credited with being the catalyst for the great strides our nation has achieved in improving water quality and protecting public health. But it is an old law, and it has not been updated through significant legislative reform in over two decades. In that time, it has become apparent that the statute's statutory structure has failed to keep pace with scientific advances, one prominent example being research on ecosystem services.

Ecosystem services flow to human communities in four streams: 1) provisioning services are commodities such as food, wood, fibre, and water; 2) regulating services moderate or control environmental conditions, such as flood control by wetlands, water purification by aquifers, and carbon sequestration by forests; 3) cultural services include recreation, education, and aesthetics; and 4) supporting services, such as nutrient cycling, soil formation, and primary production, make the other three service streams possible.

As research that has emerged and burgeoned over the past decade has shown, aquatic resources provide bountiful supplies of ecosystem services to human populations, including through groundwater recharge, storm and flood mitigation, sediment control, water purification, climate regulation, water supply, and recreation. The connections between the CWA, the central objective of which is "to restore and maintain the chemical, physical, and biological integrity of the Nation's waters," and the conservation of ecosystem services thus seem obvious and numerous, yet nowhere in the CWA are these connections made explicit. This Chapter addresses the questions of whether, where, and how those connections can be drawn so that new knowledge about ecosystem services can be integrated into decision making under the CWA.

To be sure, the CWA is not the only environmental law that has fallen behind the times in this respect. Ecologists and economists have been forging the theory and application of the ecosystem services concept since the mid-1990s, but only in the past few years has the concept begun to register in any meaningful way in federal environmental policy. Many of the environmental laws Congress passed in the 1970s have undergone little more than superficial reforms, if any, in the past twenty years, meaning new scientific concepts such as ecosystem services often find no clear home in existing statutes. This gradual scientific atrophying of environmental statutes has put tremendous pressure on administrative agencies such as the United States Environmental Protection Agency (EPA) to adapt regulatory programs to stay up to date with new knowledge and emerging policy challenges. In some cases agencies have carried out sweeping reforms at the administrative policy level, such as the broad reforms the United States Department of the Interior accomplished for the Endangered Species Act (ESA) in the 1990s. But the latitude agencies have to engage in substantive administrative reform in the absence of substantive legislative reform depends on the text and interpretations of the existing statutes on the books. Each statute thus presents its own specialised "policy space" within which an agency could, if so inclined, adjust the regulatory program through administrative reform to reflect new knowledge.

Defining Policy Space for Administrative Reform

Initiating regulatory reform in the context of stale statutory authority can be a significant challenge for an agency. Social and economic interests entrenched in and benefitted by the status quo are likely to attempt to bring political pressure on the agency to protect their interests. On the other hand, whatever conditions have prevented Congress from acting for so long in the relevant field are likely also to dampen the prospect of legislation negating the agency's regulatory reform. Much of the action in this context thus plays out in court as interests opposed to the agency's reform agenda, whatever form it takes, seek judicial review and rejection of the agency's decision as inconsistent with existing substantive and procedural requirements.

Although judicial review of agency action can take many forms and involves numerous matters for judicial consideration, the key questions in the regulatory reform context are, whether the proposed reform is consistent with the Constitution, authorised by relevant statutory authority, and compatible with the agency's existing regulations. If the answer to all three of those questions is affirmative, then all the agency need do, if even, is announce the agency's position through what is loosely described as "guidance." Although there is a point at which a substantial change in approach could be deemed to require promulgation of new legislative agency regulation, the reality is that agencies can accomplish a tremendous amount of incremental regulatory reform through guidance and other "gray law" mechanisms.

Significant regulatory innovation, however, is often going to require more significant changes to the existing regulatory regime for which mere guidance will not suffice as the sole or even primary implementation mechanism. At one extreme, agency reform action that is inconsistent with constitutional principles would require an amendment to the Constitution, which is a highly unlikely prospect. The more salient issue, therefore, is whether a proposed regulatory reform requires new statutory authorisation or only a new agency regulation.

From the agency's perspective, being able to carry out the initiative without need of new legislation may often be preferable, but it is not always clear whether the existing statute will allow it. If the new proposed regulation extends, departs from, or conflicts with prior agency regulations and practice, the agency thus must predict whether the proposal is permissible under the existing statute. In making this prediction, the agency must walk the line between two types of error: a false positive, in which the agency incorrectly concludes existing laws allow a new regulatory innovation, and a false negative, in which the agency incorrectly believes existing laws do not allow the regulatory innovation.

One of the key principles of federal administrative law provides considerable latitude for agencies in navigating between these two types of regulatory reform error. In Chevron U.S.A, Inc. v. Natural Resources Defence Council, Inc. (Chevron), the United States Supreme Court held that ambiguities in statutes within an agency's jurisdiction to administer are congressional delegations of authority to the agency to fill the statutory gap in a reasonable fashion. Filling these gaps, the Court explained, involves difficult policy choices that agencies are better equipped to make than courts. Thus, if a statute is ambiguous, and if the implementing agency's construction is reasonable, Chevron requires a federal court to defer to the agency's construction of the statute, even if the agency's reading differs from what the court believes is the best statutory interpretation.

Chevron has many nuances and has received considerable favourable and critical attention in legal and policy scholarship, but its core principle remains quite active and enforced in the courts in the context of an agency regulatory promulgation interpreting the agency's organic statutes. Indeed, more recently the Supreme Court explained that Chevron applies even when an agency is not merely extending existing policy, but also when it has completely changed directions under its statute. In National Cable & Telecommunications Association v. Brand X Internet Services (Brand X) , the Court held that if the agency adequately explains the reasons for a reversal of policy, "change is not invalidating, since the whole point of Chevron is to leave the discretion provided by the ambiguities of a statute with the implementing agency." "An initial agency interpretation is not instantly carved in stone. On the contrary, the agency ... must consider varying interpretations and the wisdom of its policy on a continuing basis," for example, in response to changed factual circumstances, or a change in administrations.

The Brand X Court went even further to hold that "a court's prior judicial construction of a statute trumps an agency construction otherwise entitled to Chevron deference only if the prior court decision holds that its construction follows from the unambiguous terms of the statute and thus leaves no room for agency discretion," and thus "only a judicial precedent holding that the statute unambiguously forecloses the agency's interpretation, and therefore contains no gap for the agency to fill, displaces a conflicting agency construction." Hence, in predicting whether a proposed regulatory reform requires new legislation or only new regulation, an agency deciding that a new regulation will suffice can take safe harbour in Chevron and Brand X Only ff it is clear from the statute that the agency has no authority to carry out the reform through regulation as a reasonable interpretation of the statute must the agency depend on legislative reform to implement the policy. The next section applies these principles to devise strategies for integrating ecosystem services into agency decision making.

Strategies for Integrating Ecosystem Services into Existing Regulatory Programs

When taken together, Chevron and Brand X provide agencies like EPA the room to experiment with regulatory innovations based on new knowledge, such as the now well-established and growing body of knowledge on ecosystem services. It is not always necessary, however, to rest on Chevron, as some statutes may clearly authorise use of ecosystem services concepts in agency decision malting. But the 2008 Farm Bill, which requires the United States Department of Agriculture to "establish technical guidelines that outline science-based methods to measure the environmental services benefits from conservation and land management activities in order to facilitate the participation of farmers, ranchers, and forest landowners in emerging environmental services markets," is at present the only such example at the federal level. At the other extreme, some statutes may make it clear that ecosystem services cannot be taken into consideration. When deciding whether to list a species under the ESA, for example, the United States Fish and Wildlife Service must take only the biological status of the species into account, and thus could not consider as a reason to list the species the economic benefits of ecosystem services that would be provided once the species and its habitat came under the statute's protection. When a statute is not clear in either of these directions, however, Chevron and Brand X are controlling.

If an agency such as EPA were to decide to integrate ecosystem services concepts into its regulatory program, therefore, the challenge would be to locate provisions in the relevant statute that can provide the platform for reasonable interpretations that using ecosystem services science in decision malting is within the scope of the agency's authority. Two different approaches seem promising in this respect. First, a statute may contain provisions suggesting that the agency can directly protect and manage natural resources for the purpose of conserving the flow of ecosystem services to human populations. For example, if a statute mandated that an agency manage or protect natural resources for, among other things, the "public welfare," one could reasonably make the argument that ecosystem services, because of their economic value and importance to human health and well being, enhance public welfare and thus maintaining or enhancing the flow of ecosystem services can be the direct focus of regulatory efforts under the statute. Declines in the flow of ecosystem services thus could be used under this "direct protection" authority to justify changes in the resource management protocol specifically for the purpose of restoring those flows.

The other approach—a fallback in the event no statutory hook credibly supports the direct protection strategy—is to locate terms in a statute that would reasonably support using ecosystem services as a criterion for determining whether the directives of the statute are being adequately fulfilled. For example, if a statute mandated that an agency manage or protect natural resources for the purposes of maintaining "environmental quality," one could reasonably argue that a credible way of determining if environmental quality is maintained is to examine trends in the flow of ecosystem services from the resource. Declines in the flow of ecosystem services thus could be used under this "performance metric" authority for deciding how to implement management changes for the resource, the incidental consequence of which would be restoring or enhancing the ecosystem services.

Using this two-pronged strategy, the challenge for agencies hoping to integrate ecosystem services into regulatory programs under statutes that do not clearly authorise or prohibit doing so is to search for provisions that reasonably can be interpreted to provide either direct protection authority or performance metric authority. While direct protection authority may be preferable for agencies hoping to establish ecosystem services as a secure focal point of regulatory policy, the advantage of the performance metric authority strategy is that it may present more flexibility for creative statutory interpretation, as Congress is less likely to put statutory sideboards on how agencies measure regulatory performance than it is on the scope of agencies' regulatory authority. The next section provides two case studies under the CWA illustrating this searching and interpretation process.

The Clean Water Act

The appropriate place to start with a holistic inquiry into the scope of the CWA, as with any statute, is with its statement of purpose—"to restore and maintain the chemical, physical, and biological integrity of the Nation's waters." One will search the statutory text in vain, however, for clues as to what "chemical, physical, and biological integrity" means. Some hints may be found in the laundry list of goals and policies Congress appended to its statement of purpose for the CWA, which include eliminating discharges of pollutants, establishing interim water quality goals for fish, wildlife, and recreation, and controlling nonpoint sources of pollution.

But these do not put meat on the bones of "integrity" and if anything broaden more than narrow the potential reach of the statute. When one turns to the operative language of the CWA, moreover, matters get no less ambiguous. As Professor Robert Adler has observed in his probing review of the meaning behind what he calls the "water quality trilogy" of chemical, physical, and biological integrity, the statute's provisions are riddled with anomalies and inconsistencies in this respect.

But Adler does find evidence of two overarching themes. First, his review of the legislative history concludes that both houses of Congress seem to have given the concept of aquatic ecosystem integrity due deliberation, indicating that the Act's opening phrase was not intended as a mere rhetorical flourish. It does appear the Act's chief sponsors in the House of Representatives and the Senate disagreed on the precise meaning of the statutory objective, as is true of many other provisions of the Act. The Senate Committee's understanding of the concept of integrity seems to suggest a return to pristine, natural ecological conditions, while the House Committee suggested that the concept of integrity implies a return to natural ecosystem structure and functions.

So there is support for the proposition that the CWA is amorphously ecological in scope, with natural conditions and functions on ecosystem scales being an intended implementation yardstick. The problem Adler detects, however, is that the regulatory teeth of the statute and of EPA's implementation for the most part have been sharpened on the chemical component of the water quality trilogy through a focus on controlling pollutants, with not enough attention to addressing the broader problem of pollution. As he observes, while progress has been made in moving toward "chemical" integrity, and while significant resources and programs have been directed at discharges of chemical pollutants, both the "physical" and "biological" integrity books in the trilogy have remained largely hortatory. Empirical evidence shows measurable gains in reducing chemical pollution, but in the thirty years since the law was passed, the overall health of the nation's freshwater aquatic ecosystems has declined dramatically.

Adler thus concludes that "neither the federal nor the state agencies charged with implementing the CWA have taken full advantage of their existing legal authority to address the physical and biological books in the water quality trilogy," and hence "it is time for EPA to revisit its virtually exclusive focus on chemical impairments to our aquatic ecosystems." That is precisely the objective in exploring how to work ecosystem services science into the statute. The two case studies that follow illustrate that the CWA presents ample opportunities for doing so, if one thinks creatively.

Section 404 Dredge and Fill Program—The Direct Protection Approach

Section 404(a) of the CWA authorises the Secretary of the Army, through the United States Anuy Corps of Engineers (Corps), to "issue permits ... for the discharge of dredged or fill material in the navigable waters of the United States at specified disposal sites." Although the Corps is the front-line regulatory agency for administering this permit program, pursuant to section 404(b)(1) of the CWA, EPA must promulgate substantive permitting standards focused on environmental factors, known as the "404(b)(1) Guidelines," which the Corps must follow when issuing permits for disposal of dredged or fill material.

Under section 404(c), EPA also may deny, or "veto," any disposal site if the discharge "will have an unacceptable adverse effect on municipal water supplies, shellfish beds and fishery areas (including spawning and breeding areas), wildlife, or recreational areas." Thus, under the section 404, and subject to specified exceptions, wetlands subject to federal jurisdiction may be filled only if the Corps grants a permit in accordance with EPA's 40400)(1) Guidelines. These permits, known ubiquitously as "404 permits," "wetland permits," or "Corps permits," have become the cornerstone for federal protection of wetland resources.

When a land development project involves filling of wetland areas regulated under section 404 of the CWA, the Corps usually requires compensatory mitigation for the loss of wetland resources as a condition of permit approval. The compensatory mitigation program suffered withering criticism for decades on a number of bases, however, one being that it failed to account for displacement of ecosystem services between fill sites and compensatory mitigation sites. Critics thus urged the agencies to incorporate ecosystem services into the array of resources directly protected under section 404.

Responding to this criticism, in 2008 the Corps and EPA jointly published final legislative regulations defining standards and procedures for authorising compensatory mitigation of impacts to aquatic resources for the Corps permits under section 404 of the Clean Water Act. Prior to the rule, the section 404 compensatory mitigation program had been administered under a mish-mash of guidance, inter-agency memoranda, and other policy documents issued over the span of seventeen years. Although motivated primarily by the need to bring the program under one coherent regulatory umbrella, the new rule also for the first time introduced ecosystem services into the mitigation program standards, requiring that "compensatory mitigation ... should be located where it is most likely to successfully replace lost ... services."

EPA and the Corps thus adopted the direct protection approach to integration of ecosystem services into the section 404 program. The question that EPA and the Corps had to contemplate, of course, was whether this is a permissible interpretation of section 404, especially in light of the facts that section 404 does not mention ecosystem services at all and that for nearly two decades EPA and Corps rules for section 404 compensatory mitigation had not mentioned ecosystem services. Enter Chevron and Brand X.

Clearly, nothing in the language of section 404 unambiguously prohibits EPA and the Corps from incorporating ecosystem services into the compensatory mitigation program, so the question under Chevron is whether doing so is a reasonable application of the statute. Piecing together such a case begins with the statute's directive that the 404(b)(1) Guidelines "shall be based upon criteria comparable to the criteria applicable to the territorial seas, the contiguous zone, and the ocean under section 1343(c) of this title." That provision in turn specifies that the guidelines address the following criteria, with emphases added to point to the hooks upon which to hang ecosystem services:

(A) the effect of disposal of pollutants on human health or welfare, including but not limited to plankton, fish, shellfish, wildlife, shorelines, and beaches; (B) the effect of disposal of pollutants on marine life including the transfer, concentration, and dispersal of pollutants or their byproducts through biological, physical, and chemical processes; changes in marine ecosystem diversity, productivity, and stability, and species and community population changes; (C) the effect of disposal, of pollutants on esthetic, recreation, and economic values...

Through this incorporation of regulatory goals, including human health and welfare, marine ecosystem diversity, productivity, and stability, and recreation and economic values, section 404 gives EPA and the Corps ample room to include ecosystem services among the resources that the program is designed to protect. Ecosystem services from wetland resources support human health and weffare; the diversity, stability, and productivity of marine ecosystems support their capacity to supply those ecosystem services; and providing esthetic, recreation, and other economic values is the policy objective of incorporating ecosystem services knowledge into decision making. Given the strength of these connections, it would be futile under Chevron to argue that EPA and the Corps misconstrued section 404 and unreasonably incorporated ecosystem services as a protected resource, and Brand X dispenses with any objection that the abrupt change in policy is impermissible.

Section 303 Water Quality Standards and TMDL Program—The Performance Metric Approach

Whereas section 404 expressly focuses agency decision making on impacts to aquatic ecosystem health, thus providing the base of support for the direct protection strategy for incorporating ecosystem services into decision making, the section 303 program presents much less opportunity for pursuing that strategy. The performance metric approach, however, seems well suited to section 303.

The section 303 water quality standards and TMDL program work in tandem with the permitting provisions of the CWA found in section 402, which authorises EPA to issue permits "for the discharge of any pollutant" other than discharges covered in provisions such as section 404. Discharge is defined in the CWA so as to limit section 402 to pollutants emitted from "point sources," which are defined as confined and discrete conveyances, such as pipes and ditches. For such discharges, section 402 sets up an extensive technology-based effluent control standards and a permitting program known as the National Pollutant Discharge Elimination System (NPDES).

Section 303 enters the picture when the technology-based effluent limits imposed on NPDES dischargers under section 402, even with full compliance, are inadequate to meet water quality goals for specific water bodies. Section 303(c) requires states to prepare and present for EPA approval water quality standards consisting of "the designated uses of the navigable waters involved and the water quality criteria for such waters based upon such uses." To assist states in this function, section 304 requires EPA to develop guidelines for establishing water quality criteria "accurately reflecting the latest scientific knowledge."

Section 303(d)(1)(A), in turn, requires states to identify waters for which the technology-based effluent limitations imposed through section 402 permits are not adequate to attain an applicable water quality standard, and section 303(d)(1)(C) requires states to establish and implement for such listed waters the "total maximum daily load ... for ... pollutants" as a means of reducing discharges to levels that will attain the water quality standard.

Nothing in these provisions overtly addresses ecosystem services science one way or the other; hence, as with section 404, EPA would have to dig deeper into them to search for and interpret authority for the agency to incorporate ecosystem services science into the section 303 program. At the threshold level of such an inquiry, section 303 divides into two distinct components—the section 303(c) water quality standards component and the section 303(d) TMDL and load allocation component. The load allocation component is the regulatory branch of section 303 and thus where one would search for direct protection authority. But section 303(d) is a set of dry technical provisions devoid of opportunities for creative interpretation along these lines. All the policy space in section 303 lies instead in the water quality standards component, which is fundamentally about establishing the criteria for assessing the performance of technology-based effluent standards imposed on dischargers under section 402 NPDES permits. Hence, if there is ecosystem services gold to be mined in the section 303 program, it will come by searching for provisions in section 303(c) that can reasonably be interpreted to provide performance metric authority.

The first such provision is a potential bonanza for the performance metric approach: section 303(c)(2)(A) requires that water quality standards "shall be such as to protect the public health or welfare, enhance the quality of water and serve the purposes of this chapter." The provision goes on to require that the standards take into consideration the "use and value" of the water body for, among other things, "public water supplies, propagation of fish and wildlife, recreational purposes, and agricultural, industrial, and other purposes." Furthermore, EPA's authority under section 304 to develop guidelines the states are to use to designate water quality standards under section 303(c) requires that EPA consider "all identifiable effects on health and welfare, including, but not limited to ... shorelines, beaches, esthetics, and recreation which may be expected from the presence of pollutants in any body of water," as well as "the effects of pollutants on biological community diversity, productivity, and stability."

As with section 404 (through its incorporation of section 403), these provisions contain strong connections to ecosystem services, opening the door wide to incorporating ecosystem services science broadly into water quality standards formulation. By using ecosystem services such as groundwater recharge and flood control to describe designated uses of water bodies, and by specifying levels of ecosystem service flows to human populations as water quality criteria, EPA could move the section 303 program closer to Adler's vision of the trilogy of physical, chemical, and biological water quality goals. For example, if adequate capacity for flood control were described as a water quality standard for a water body, pollutants degrading the aquatic vegetation contributing to that flood control capacity could trigger an impairment finding and the appropriate responses under the TMDL component of section 303. The language of sections 303 and 304 surely would support doing so under Chevron, and the fact that the agency has not done so in the past would present no obstacle given Brand X.

The hitch, as noted above, is that as a performance metric program, the water quality standards component of section 303 depends entirely on the load allocation component of section 303(d) for its regulatory implementation. Also, although nonpoint sources such as agriculture, urban runoff, and unspecified sources are significant contributors to water quality impairment, EPA is limited under 303(d) in its ultimate reach to regulating discharge of pollutants from point sources subject to section 402 NPDES permits. Nevertheless, states must identify a water body that fails to meet water quality standards as impaired even if nonpoint sources are contributing to the impairment, and the load allocation must include such nonpoint sources. If states choose not to regulate nonpoint sources and thereby fail adequately to implement the TMDL program, EPA can withdraw federal grant money from the state.

Hence, although it is true that "by limiting the effective control mechanisms to total maximum daily loads of pollutants, Congress included in section 303(d) no direct mechanisms to redress other sources of pollution," there is still good reason to include ecosystem services as one of the performance metrics of water quality standards. EPA recently has claimed to be committed to modernising the section 303 program by busting the water quality standards regime out of its chemical focus and moving it closer to Adler's trilogy vision. Using ecosystem services science to do so will allow EPA to regulate point source discharges in water bodies with impaired ecosystem services, will call attention to the contribution of nonpoint sources to ecosystem services impairment, and will put pressure on states to address those nonpoint sources. Chevron and Brand X do not allow EPA to overcome the structural limits of regulatory authority under section 303(d), but they could pave the way for EPA to use sections 303(c) and 304 as a platform for making ecosystem services an important performance metric of the section 303 program.

Speaking about one of the CWA's siblings, the Clean Air Act, the Supreme Court recently observed that Congress understood when it designed the statute "that without regulatory flexibility, changing circumstances and scientific developments would soon render the Clean Air Act obsolete," and hence "the broad language of [the statute] reflects an intentional effort to confer the flexibility necessary to forestall such obsolescence." This principle, of course, is not limited to the Clean Air Act—the CWA embodies the very essence of employing broad language to impart the flexibility needed to incorporate scientific developments. Indeed, Congress made doing so an explicit command.

The science of ecosystem services has emerged as a powerful organising principle of interdisciplinary ecological, economic, and social research, and has begun to take hold in policy formulation. Within EPA, for example, the Office of Research and Development's Ecosystem Services Research Program is transforming the way we account for the type, quality, and magnitude of nature's goods and services so that they can be considered in environmental management decisions. The research is providing the data, methods, models, and tools needed by states, communities, and tribes to understand the cost and benefits of using ecosystem services.

Clearly this is the type of scientific advancement that Congress contemplated agencies would incorporate into administration of environmental statutes such as the CWA. The judicial review principles embodied in Chevron and Brand X facilitate this incorporation process by allowing agencies to adapt statutory provisions to new knowledge and by relieving agencies of the concern that change is impermissible because "that's not how we've done it."

As Adler's work emphasizes, the narrow chemical pollutant focus EPA has taken with its CWA authority is neither demanded by the statute nor consistent with contemporary scientific perspectives on aquatic ecosystem health and integrity. Incorporating ecosystem services science into the CWA programs as broadly as possible will be one important component of moving the statute forward in this respect. This Chapter has laid out the strategies for so incorporating ecosystem services science and illustrated their application with two cases studies of discrete CWA programs.

Agencies like EPA and the Corps, if they are committed to staying scientifically relevant, need not and should not walt for Congress to graft new science into statutes. Instead, agencies should scour statutes like the CWA and other environmental laws for opportunities to interpret the existence of direct protection and performance metric authority as leverage points for incorporating new science into evolving regulatory programs. To be sure, the science of ecosystem services should not be sitting on the shelf waiting to be dropped into the CWA and other environmental statutes. Further research on ecosystem services tailored to regulatory programs such as section 404 and section 303 will be needed, which fortunately has begun at EPA, and numerous policy design questions must be addressed. But by no means should the existing set of environmental statutes themselves be seen as insurmountable obstacles simply because they were designed before the concept of ecosystem services took hold in the scientific community. Fitting new science into old laws will take political will and some creative interpretations of stale laws, but it will be necessary if our environmental statutes are to remain relevant and effective. As this Chapter has shown, the new science of ecosystem services presents just such an opportunity.

Water Law and Associated Stories

In some respects, the Pacific Northwest as a region has been defined by what it is not: it is not as arid as the intermountain West. But it is a mistake to think that water in the Northwest is not a scarce resource, subject to increasing demands for a variety of consumptive and nonconsumptive purposes. The 1990s have made evident that Northwest water supplies are insufficient to meet regional needs, as perhaps best illustrated by the listing of Snake River salmon for protection under the Endangered Species Act (ESA). Thus, the myth of abundance has been only recently exposed. In truth, a student of water law in the Northwest must confront a number of myths that inhibit a full understanding of the framework within which rights to the region's most precious natural resource are allocated.

Northwest Water Law Myths

Myth No. 1: The Northwest is Humid

It is true that the Northwest is home to the largest remaining temperate rain forest in North America, and climatic conditions here are wet enough to provide ideal conditions for the growth of Douglas fir trees. But heavy moisture characterises only the Pacific slope of the Cascade Mountains, which comprises but a small fraction of the four states of the Pacific Northwest. Roughly two-thirds of Oregon and Washington lie east of the Cascades as, of course, do all of Idaho and Montana. East of the Cascades aridity rules. While precipitation in the Olympic peninsula northwest of Seattle can average over one hundred inches a year, Seattle's average precipitation averages only about forty inches, and just one hundred miles east of Seattle, on the east side of the Cascades, annual precipitation averages less than nine inches. Moreover, virtually all of the region's rainfall occurs between November and April, before the summer irrigation season begins. During the summer, in the majority of the region lying east of the Cascades, the Northwest is arid. Because most of the people in the Northwest live along the narrow ribbon on the Pacific slope of the Cascades, the region's geography has masked this essential aridity. Those who wish to understand the reality of the critical importance of water in the Northwest must look behind the mask.

Myth No. 2: Salmon Divides the Region into Lower and Upper Basin Constituencies

The great river that drains all of the states of the Pacific Northwest, the Columbia, and its principal tributary, the Snake, unite the region as no other resource does. Over the past half-century, the region constructed (mostly with substantial federal subsidies) the largest interconnected hydroelectric system in the world. Hydropower gave the Northwest the cheapest electric rates in the nation, brought electricity to rural areas east of the Cascades, and lured to the region electric-intensive industries, such as aluminium plants, that diversified the regional economy. Unfortunately, the damming of the Columbia River crippled what were once the world's largest salmon runs. For the past twenty years, coincident with the completion of the last mainstem federal dam, the region has been preoccupied with salmon restoration. That concern produced the innovative fish and wildlife provisions of the 1980 Northwest Power Act, which in turn produced the largest biological restoration program on the planet. Unfortunately, the ineffectiveness of that program prompted listing of Snake River sockeye and chinook salmon runs for ESA protection. The ESA listing' led to moderate changes in Columbia River flows, increasing flows somewhat in the spring and summer when juvenile salmon migrate to the ocean. Salmon advocates have protested that these flow increases are too small to restore the endangered runs.

While some debate the adequacy of salmon flows, others question the efficacy of flow increases at all. Since Idaho Governor Cecil Andrus (D) left office, Idaho has resisted increased Snake River flows where those flows would be the result of storage reservoir drawdowns in Idaho. Although it is plain that Idaho salmon require Idaho water, Idaho officials have questioned the scientific basis of increased river flows, and the state has even made water transfers more difficult where the result is to send water downstream, out-of-state.

The upshot of Idaho's reluctance to support river flows for salmon restoration has been to create the perception that there is a divide between the upper basin states and the lower basin states. Such a division would, for example, produce a stalemate on the Northwest Power Planning Council, where upper and lower basin states are represented equally. While there is no specific requirement that the Council increase water flows to facilitate salmon migration, there may be limits on unilateral actions inconsistent with salmon restoration that an upstream state such as Idaho can take under state law. For instance, if Idaho were to deny a water right transfer aimed at boosting Snake River salmon flows, that denial could be subject to a Commerce Clause challenge for discriminating against out-of-state users.

Such a denial would also seem to violate Idaho's affirmative conservation duty under the equitable apportionment doctrine, which the U.S. Supreme Court has ruled applies to salmon. In so ruling, the Court concluded: "At the root of the [equitable apportionment] doctrine is the same principle that animates many of the Court's Commerce Clause cases: a State may not preserve solely for its own inhabitants natural resources located within its borders." As a result, if Idaho were to deny a water transfer whose purpose was to improve flows in the Snake River for the benefit of salmon on the ground that it was "conserving" water for in-state use, that denial could violate both the Commerce Clause and the equitable apportionment doctrine. As the Court concluded, "States have an affirmative duty... to take reasonable steps to conserve and even to augment the natural resources within their borders for the benefit of other States."

Thus, even if the political winds seem to be blowing the upper and lower basin states apart on salmon recovery issues, the courts may insist that upper basin states not hoard water necessary for effective salmon migration. If so, it will have been the least representative branch of government that refuses to allow us to escape the reality that the salmon and their migrations are the fundamental ties that bind us together as Northwesterners.

Myth No. 3: The Market Allocates Western Water Rights

One of the myths propounded by advocates of the prior appropriation system of water rights is that it is a market-driven system. According to this myth, although Western water law rewards the first diverter (who may or may not be the most efficient user) with property rights, the prior appropriation doctrine's great virtue is that it allows alienability of rights, so that water may flow to the user who values it most, measured by willingness to pay. For some advocates of rules promoting efficient use of natural resources, Western water law's emphasis on private rights and market-based transfers provides a paradigmatic illustration of the operation of the Coase Theorem's contention that market transfers will produce economic efficiency.

But the reality of water allocation under the prior appropriation doctrine is that the market operates with frequency only in some states, and then usually only to transfer one diversionary use to another. The "no injury" rule to third parties has worked to foreclose many transfers, as has the failure of the seller to demonstrate full use of his water right. In the Snake Basin, efforts to purchase water from agricultural users to increase river flows to improve salmon migration have met with considerable local resistance.

This resistance is not based on market principles, such as sellers demanding a higher price to forego diversions necessary for agricultural crops. Instead, the resistance is due to local community opposition to market transfers where those transfers would take land out of agricultural production, with attendant erosion of tax base and adverse local economic effects. This resistance to the potential consequences of a free market caused the 1995 Oregon House to pass a bill that, had it not been killed in the Oregon Senate, would have banned market transfers of agricultural water for the purpose of increasing instream flows.

In truth, then, the market is far from the overriding force in allocating Western water rights. Where the market conflicts with local community values, water market transactions are regularly thwarted, a reminder of the strong communal interest in Western water.

Myth No. 4. Appropriators Are Entitled to a Fixed Quantity of Water

Although it is hornbook law that the measure of a water right is beneficial use without waste, many diverters continue to believe that the quantity of water they may legally divert is a constitutionally protected property right. This myth has been perpetuated by the courts, which have only rarely restrained historic uses by finding customary methods of diversion to be wasteful. Nevertheless, prior appropriation protects only nonwasteful beneficial uses, not historic diversions. As modern irrigation systems reduce evaporation and seepage losses in conveying water to crops and increase the percentage of diverted water that can be consumed by the crop, outmoded carriage and application systems can be declared wasteful, and therefore not within the diverter's water right.

The concept of nonwasteful beneficial use thus has the potential to demand cessation of historic wasteful practices and adoption of modern conservation measures. The Washington Supreme Court recently supplied an example of how stricter application of the beneficial use and waste doctrines can help modernise Western water law. In Department of Ecology v. Grimes, the court upheld a referee's decision that refused to recognise a diverter's historically wasteful irrigation practices as the measure of his water right. According to the court, while customary irrigation practices are a factor in determining what constitutes waste, they are not determinative. Thus, the wasteful diverter had no right to constitutional compensation as a consequence of the referee's decision limiting his right to divert water.

Proper understanding of the scope of the property right in water under the beneficial use and waste doctrines not only means that state efforts to limit wasteful diversionary practices will not run afoul of the Constitution, it also makes less revolutionary the application of public trust principles to Western water rights. After all, the essence of the public trust is accommodating both instream and diversionary needs; rigorous application of the beneficial use and waste doctrines could free up diverted water for instream (public trust) purposes. The Idaho Supreme Court's recent affirmation that the public trust burdens all water rights in that state gives the public some hope that Idaho courts will insist upon a reconciliation between instream and diversionary uses, which will necessarily result in curtailment of wasteful diversions. Similar results may be occasioned by implementation of the Endangered Species Act and the Clean Water Act.

Myth No. 5: Hydropower Law is Not Water Law

Over the next fifteen years, the Federal Energy Regulatory Commission (FERC) must consider relicensing some two hundred nonfederal dams in the Northwest under the terms of the Federal Power Act. The outcome of these procedures may have a substantial effect on Northwest water flows, so it is a mistake to think that these federal adjudications are not an integral part of water law. In fact, the courts have ruled that a federal hydroelectric licenses can preempt state water law. The most significant FERC relicensing proceeding on the regional horizon is Idaho Power Company's three-dam Hells Canyon complex.

Authorisation of these projects in the 1960s extirpated the salmon runs of the middle Snake Basin, despite Idaho Power Company's assurances that a trapping and hauling program around the dams would preserve the runs. Today the Hells Canyon complex is an important spigot of storage water, essential for downstream salmon flows. The relicensing process could produce higher flows to aid salmon migration in the spring and summer months and might even call for a reintroduction of salmon in the Snake Basin above the three-dam complex. A similar reintroduction of salmon in the upper Deschutes Basin in central Oregon is under consideration in connection with the relicensing of Portland General Electric's Pelton and Round Butte Dams.

The U.S. Supreme Court's recent decision in the Dosewallips River case has helped to expose as myth the notion that hydroelectric licensing is separate from water quality law. In ruling that the state could impose minimum flow conditions on a FERC-licensed project through section 401 of the Clean Water Act, Justice O'Connor's opinion for the Court referred to an attempted separation of water quantity concerns (as reflected in the FERC licensing process) and water quality concerns (under the Clean Water Act) as an "artificial distinction." She noted that "in many cases, water quantity is closely related to water quality; a sufficient lowering of the water quantity in a body of water could destroy all of its designated uses, be it for drinking water, recreation, navigation or, as here, as a fishery." Advocates of water law reform should adopt Justice O'Connor's holistic view of the water resource and view the FERC relicensing process as a central element in restoring water flows in many Northwest river basins.

Myth No. 6: Water Law is Unconnected to Watershed Protection

The Supreme Court's recognition of the inseparable relationship between water quantity and water quality in the Dosewallips case calls into question Western water law's narrow focus on the stream channel. In the rural West, water quality is largely a function of upland land management, especially federal land management. The link between water law and watershed management has been noted in the leading public land treatise, which observes that "excessive diversions can harm watershed values, and watershed protection can assure sustained yield of high quality waters for downstream use." Yet legal recognition of watershed values is in its infancy, and even the definition of what constitutes a watershed is unclear.

The Ninth Circuit has initiated judicial recognition of watershed protection by recognising the connection between federal land management activities and state water quality standards in a series of recent cases. In holding that citizens may enforce state water quality standards against federal timber harvests, the courts have begun to forge a "reluctant marriage" between watershed concerns and federal land management. But these cases depend on clear violations of enforceable state water quality standards, which may prove to be more the exception than the rule. Perhaps a more promising means of enforcing this reluctant marriage, at least on national forest lands, is through the National Forest Management Act's provisions promising that forest plans must protect water quality and that clearcutting will not injure watershed values. One court has remanded a forest plan for failure to assure watershed protection. It is also possible that public land managers could foster watershed management through their authority to condition rights-of-way necessary for water diversions upon provision of "bypass flows" to protect downstream watershed conditions.

Perhaps the paradigm example of managing federal lands to protect watersheds is the Clinton Administration's efforts to redirect Northwest public land timber harvests to protect species—like the northern spotted owl, the marbled murrelet, and the Snake River salmon—listed under the Endangered Species Act. However, the Administration's plans have been crippled by enactment of the "timber salvage" rider as a part of the 1995 Rescissions Act, Judge Michael Hogan's sweeping interpretations of the scope of that rider, and a congressional threat to defund ecosystem planning east of the Cascades. Yet at the same time that Congress is threatening to end the federal government's involvement in watershed management, watershed planning on the local level appears to be on the rise. The logic of managing lands and waters on a watershed basis, first recognised by John Wesley Powell over a century ago, may be so persuasive that even the 104th Congress cannot stop it.

Myth No. 7: Indian Treaty Fishing Rights are Not Water Law

Federal reserved water rights have long been a source of controversy throughout the West because they do not depend on state law concepts such as diversion or beneficial use, and because reservations were often established in the nineteenth century, giving them valuable early priority dates. States have gained some control over reserved rights through broad judicial interpretations of the McCarran Amendment, a congressional waiver of sovereign immunity that allows federal reserved water rights to be adjudicated in state comprehensive basin water adjudications. Nevertheless, despite predictably narrow state court interpretations, Indian reservations have secured large amounts of reserved rights. Until recently, however, courts have based their recognition of reserved water rights largely on agricultural purposes of reservations. Yet many Indian reservations in the Northwest have both fishing and agricultural purposes, and under the Stevens Treaties the signatory tribes also possess off-reservation fishing rights. The scope of water reserved to carry out these purposes remains largely unsettled, but the amount could be quite large.

Among the more prominent interpretations of the scope of the reserved water for fishing is the federal district court decision in United States v. Anderson, which concluded the Spokane Tribe had sufficient water reserved in Chamokane Creek to maintain water temperatures cold enough to promote fish spawning, and the Ninth Circuit decision in Colville Confederated Tribes v. Walton, which ruled that the reserved water in Omak Lake was the amount necessary to maintain a replacement fishery for the Colville Tribes. In another decision, the Ninth Circuit in United States v. Adair distinguished the Klamath Tribe's reserved water for fishing from the Tribe's reserved water for agriculture, ruling that the former, with a priority date of "time immemorial," takes precedent over the latter.

As to the scope of the reserved water right for fishing, the court determined that the treaty did not reserve sufficient water to restore a treaty-time "wilderness servitude," and instead reserved only sufficient water to maintain fishing "as currently exercised." Similarly, the Washington Supreme Court, in Department of Ecology v. Yakima Reservation Irrigation District, concluded that because the scope of the Yakama Nation's reserved water for fish had been diminished by post-treaty congressional policies favouring irrigation in the basin and an Indian Claims Commission settlement, the scope of the right was limited to "the minimum instream flow of the river necessary to maintain anadromous fish... [under] annual prevailing conditions."

The latter two results could be interpreted to mean that the amount of reserved water for fish is not sufficient to promote restoration of damaged fish runs, a result that would be contrary to the results in Anderson and Colville Confederated Tribes. However, on remand in the Yakima case, the trial judge employed the "maintain anadromous fish under annual prevailing conditions" standard to order storage releases to increase flows in tributary streams and to increase "flushing flows" in the mainstem Yakima River. Thus, some restoration of the Yakama Nation's fish runs is underway. The amount of restoration to which fishing tribes are entitled is not clear, but the answer is more likely to be a function of the degree to which Congress has taken action in direct conflict with treaty promises to continue to fish than a reflection of the existing condition of the fishery.

Uncertainties associated with the scope of the reserved water right for fish, coupled with the clear temporal seniority of the tribes' fishing right over virtually all other water rights, has induced some competing water users to seek to negotiate agreements with the tribes. Over twenty years ago, federal water managers promised not to "impair or destroy" the fishing rights of the Umatilla Reservation tribes. More recently, irrigators in the Dungeness Basin agreed to curtail water diversions to help restore salmon runs in which the Jamestown S'Klallam Tribe has treaty rights. Settlements of treaty rights to reserved water are proliferating throughout the West, and there is no reason to think that there will not be more settlements in the near future involving treaty rights to fish.

The Northwest Water Law & Policy Project, a foundation-supported think tank studying Northwest water law and policy issues, sponsored its first spring conference in May 1995. Earlier, the project sponsored a fall conference on Columbia River governance, which was subsequently reprinted in Environmental Law. At the spring conference, the project released several papers for public review and comment, many of which are published in this symposium. Others are available from the project, as is the project's quarterly newsletter. David Getches's keynote address blames water law and policy for much of the destruction of the Columbia River's fabled salmon runs, which he refers to as "the buffalo of the Northwest Indians." He argues that water law and policy, however, may be redeemed through evolution of the concept of beneficial use to become "an engine of the public interest," through modifying dams and diversions to allow them to be managed for ecosystem benefits, and through the rise of governance on the watershed level, which may allow local citizens to devise solutions to water conflicts that are fine-tuned to specific places and real people. Through watershed planning, Professor Getches implies we may be able to recreate conditions that will foster the kind of civic virtue the nation's founders thought essential to pursuit of the public interest.

Dar Crammond explores leasing water for instream flow enhancement as a means to improve ecosystem functions on the watershed level. His article considers a variety of potential water sources for leases, examines federal and state laws in the Northwest affecting water leases, including the unsettled question of whether private parties may control leased instream rights, and suggests strategies for valuing and negotiating leases for instream purposes. His article concludes that the glacial rule of instream flow leasing in the Northwest is due to the novelty of the concept, limited agency budgets inhibiting development of efficient leasing programs, and the uncertainty over whether individuals may hold property rights in instream flows.

Reed Benson evaluates ongoing watershed planning in the Northwest in terms of its capability to restore water flows for instream uses such as salmon spawning and migration. Surveying watershed management activities throughout the Pacific Northwest, Benson concludes that most plans focus on land use and riparian measures, not on protecting and restoring instream flows. He argues for including streamflow measures in watershed plans but cautions that doing so will require overcoming inflexible state water laws and political hostility to governmental management of natural resources.

Shauna Whidden considers the future of the last free-flowing stretch of the Columbia River, the Hanford Reach. The Hanford Reach provides spawning habitat to the largest remaining wild populations of Columbia River salmon, due largely to the fact that the nearby Hanford Nuclear Reservation blocked development of water projects. Ms. Whidden explains that now that the federal government has changed the reservation's purpose from Cold War-driven nuclear arms production to environmental restoration, the lands adjacent to the Hanford Reach may be preserved as a national wildlife refuge or sold to private parties for agricultural development. She contends that Congress should adopt the recommendations of a recent National Park Service study by designating the Hanford Reach as a wild and scenic river and preserving the adjacent lands in federal ownership by designating them a national wildlife refuge.

Finally, Joy Ellis considers the moratoria on water diversions from the Columbia River that Idaho, Oregon, and Washington put into place in response to the endangered species listings of Snake River salmon species. She examines the effectiveness of each of the state laws, evaluating exceptions to the moratoria and the exceptions' applicability to tributary streams. She explains that the most effective of the moratoria, Idaho's, is actually a drought measure aimed at facilitating groundwater recharge, not salmon recovery. The chapter explores the role of the Endangered Species Act in exerting control over the states' moratoria and concludes with suggestions on how to improve the effectiveness of moratoria, including a proposal that state agencies subordinate any new water rights to later developed instream rights.

The Adoption of Water Conservation

Proponents of environmental conservation policies face significant costs and risks in the political arena since effective policies typically regulate or redistribute social resources. As students of the policy process have long recognised, regulatory and redistributive policies are likely to produce political conflict. Adoption of new environmental conservation policies, thus, requires policy entrepreneurs who are willing to bear the risks and costs of innovation. Despite their popular depiction as risk averse, public administrators are identified as entrepreneurial drivers of environmental policy innovation in several recent studies. In light of the risks and costs of policy entrepreneurship, why do bureaucrats pursue environmental policy innovations?

This chapter investigates the ways that bureaucratic professions and their labour markets affect the diffusion of environmental conservation policy in the

United States. The bureaucrats who serve American local governments are, by and large, career professionals engaged in policy networks with other professionals. Professional networks can be sources of policy ideas and political resources and so can help explain the diffusion of policy innovations from professions to government agencies. That bureaucrats sometimes emerge as environmental policy entrepreneurs also is well known, as noted above. We know little, however, about why bureaucrats in some agencies are more likely than others to drive environmental policy innovations. Why do some administrators introduce policies from their professions into the agencies that they serve, whereas others do not? When is a bureaucrat likely to push an environmental policy favoured by her profession? This chapter argues that bureaucratic job mobility increases the likelihood that environmental protection policies will diffuse from professions to governments. When an agency head arrives from outside the government she serves, she carries both a reputation and mandate for professional innovation. Conditions are less conducive to professional innovation when an agency head is promoted from within. The result is mobility-contingent professionalism, for the priorities of an administrator's profession are more likely to become manifest in policy when she arrives from outside than when she is promoted from within an agency. In short, in matters of environmental conservation policy, professionalism matters more for the mobile bureaucrat. Mobility-contingent professionalism offers a new way to think about the relationship between professions and the diffusion of innovations across governments.

The present study tests this theory of mobility-contingent professionalism by modelling the adoption of conservation-oriented water rate structures in American public utilities. In addition to being a redistributive environmental protection policy, conservation-oriented water rates are also highly salient and generally favoured by the water utility profession today. After reviewing the conceptual issues at hand, I describe how career paths favour more or less diffusion of environmental protection policy through the selection of candidates for administrative jobs. I briefly relate an illustrative case of a utility that recently adopted water conservation rates under the leadership of a professionally active, mobile professional manager. Drawing data from an original survey of water utility managers, I then present a pair of statistical models that demonstrate the ways in which career paths and professional identity affect conservation policy adoption. Finally, I summarise the results of the analyses and highlight their theoretical and normative implications for the study of policy diffusion generally and environmental policy particularly.

Environmental Policy Entrepreneurs

Scholars have sought to understand the process of policy innovation in government since Walker's (1969) seminal article on the spread of policies from state to state. Innovation in these diffusion studies means the introduction of a policy new to the government adopting it. Studies of innovation have isolated a number of social, economic, and institutional correlates of policy diffusion and so have identified several conditions that promote or inhibit the spread of innovations from one government to another. Over the past decade, studies of policy diffusion and innovation have begun to pay attention to the specific mechanisms by which diffusion occurs and to the people whose political decisions cause innovation. These studies identify policy entrepreneurs as important causes of policy innovation. Policy entrepreneurs are individuals who recognise latent demand for new policies and then expend resources and bear political risks to drive policy innovation.

Policy entrepreneurs figure especially prominently in recent studies of the diffusion of environmental protection policies. Elected officials seize entrepreneurial opportunities to advance environmental policies at the local, state, national, and international levels. Bulkeley and Betsill's (2003) study of local governments' responses to climate change identifies entrepreneurial bureaucrats as drivers of policy innovation. Similarly, Rabe (1999, 2004) finds that professional administrators frequently emerge as entrepreneurs who drive innovations in pollution control and climate change policy. That entrepreneurs are significant in environmental protection politics is unsurprising since environmental policies are regulatory and/or redistributive. Regulatory policies control the use of private resources, whereas redistributional policies affect the distribution of resources among individuals. Regulatory and redistributional policies are prone to controversy and so are costly and risky to pursue.

Balla (2001) and Mintrom and Vergari (1998) find that involvement in professional policy networks is related to the diffusion of policy innovations. With respect to environmental policies in particular, Rabe (1999) finds that entrepreneurial administrators borrow innovative policy ideas from "policy communities" related to their professions. In these studies, professional organisations and related issue networks are resources that policy entrepreneurs draw upon in pursuit of innovation.

Several accounts of administrative behaviour suggest that the process of professional accreditation (through formal education, apprenticeship, and so forth) imbues individuals with the ethics of their professions. Steeped in the cultures of their professions, administrators come to understand good and bad policy according to the conventions of their professional peers, goes the argument; this socialisation process causes administrators to be, as Brehm and Gates put it, "principled agents." An administrator's preference for the policies favoured by his profession may follow from his very identity as a professional, gained through years in college, graduate school, and service in the ranks of his fellow professionals.

Left unanswered is why professional bureaucrats-turned-entrepreneurs bother pursuing innovation in the first place. Invoking Mohr's (1969) theory of organisational innovation, Berry and Berry (1990, 399) identify two factors that increase the probability of innovation in organisation: "(1) the motivation to innovate and (2) the availability of resources for overcoming obstacles" to innovation. Theories of policy innovation to date have had much to say about the latter but scarcely anything systematic about the former. Why do policy entrepreneurs—especially bureaucrats, whose jobs may be at stake—bear the costs and risks of innovation? As Tews, Busch, and Jorgens (2003, 593) observe, "...the question about the concrete motivations of policy makers to adopt environmental policy innovations is still unanswered."

The approach of Berry and Berry (1990) to the motivational question is typical of policy diffusion studies: they conceptualise the motivation to innovate in terms of a need or problem that demands a policy solution. Since the policy they study is adoption of state lotteries, Berry and Berry use a state's fiscal condition as an indicator of its motivation to innovate. Similarly, the study by Zahran et al. (2008) of local climate change policy uses vulnerability to natural disasters as a measure of motivation to adopt climate change policy. This approach is appropriate when the objective of the study is to find economic, institutional, and social correlates of innovation. But when examining a diffusion-by-policy-entrepreneur model, to conceptualise "motivation to innovate" as a public need is to conflate individual goals with organisational goals. As organisation theorists have long recognised, individuals do not necessarily share the goals of the organisations in which they work. It is one thing for a local utility to suffer a drought; it is another for a bureaucrat to assume costs and risks in pursuit of a controversial policy to address the crisis. On its own, need for a policy is an inadequate explanation for the emergence of policy entrepreneurs.

As Kingdon (2003) described them, the defining characteristic of policy entrepreneurs is "...their willingness to invest their resources—time, energy, reputation, and sometimes money—in the hope of a future return" (123, italics added). Kingdon suggests that entrepreneurs may champion policies that they believe are important, or that they simply enjoy the act of political engagement in itself. At first blush, professional socialisation would seem to underlie the "willingness to invest resources" and to offer the "future return" that Kingdon specifies.

But professional identity alone is theoretically problematic as an explanation for bureaucratic policy entrepreneurs. Professional identity and the prestige associated with a profession are collective goods that offer free-riding opportunities for professionals. A bureaucrat may enjoy good standing in her profession and the prestige and perquisites of office without incurring the costs of championing her profession's favoured causes. It is not clear why administrators would take significant political risks in the name of professionalism alone. Moreover, professional socialisation fails to account for variation in administrators' pursuit of professionally sanctioned goals. In an age when virtually every government agency of any substantial size is staffed by professionals, why do professional innovations emerge in some agencies and not in others?

Theory

Kingdon (2003, 123) identifies a third potential motivation that might spur policy entrepreneurs to innovate: "promoting one's personal career." This study seeks to explain the diffusion of water conservation rates from professions to local governments by focusing on bureaucrats as policy entrepreneurs. Agency heads are well positioned to drive innovation, and their professions provide the "raw materials" of entrepreneurship in the form of new policy ideas. But the incentive to pursue professionally sanctioned innovations lies in the professional labour market, I argue. Professions may have preferred policies, and administrators may be socialised and identify as professionals. But job mobility—the movement of bureaucratic professionals from one government to another—creates conditions under which bureaucrats' professional sensibilities are likely to become manifest in policy. Like most jobs in a market economy, bureaucratic jobs are temporary matches of individuals with employers. But individuals and agencies do not latch on to one another at random, like so many atoms in Brownian motion. Governments' selection processes and individuals' adaptation to those processes shape bureaucratic behaviour and, ultimately, government policy.

Selection

When elected officials hire professional administrators, they are usually laypersons hiring experts. So as in most hiring situations, qualifications and reputation are important factors for elected officials selecting agency heads. But the procedures and criteria applied in selection of agency heads depend on whether bureaucrats are promoted from within the agency or hired from outside.

Professional credentials and reputation are prominent selection criteria for governments hiring bureaucrats from outside. Elected officials rely heavily on the advice of other professionals when vetting and selecting candidates. Local governments hiring agency executives from outside their organisations typically hire executive search firms or consultants specialising in the professions at hand. These search consultants are usually themselves former administrative professionals, and they are influential in framing issues and establishing evaluative criteria. With little knowledge about the candidates, elected officials rely heavily on the advice of their search consultants and the candidates' credentials and reputations. Overall, in evaluating external candidates' reputations, local governments hiring from outside scrutinize applicants through the prism of professionalism. If a profession generally favours conservation policies, then governments are likely to favour professionals with a demonstrated interest in or success with such policies.

Few or none of these selection processes apply when organisations promote internal candidates as a matter of policy. Governments promoting from within the organisation are familiar with their candidates and select an agency head with whom they are comfortable. Organisations with a standing practice of hiring executives from within the organisation almost certainly have fewer candidates for the job—perhaps only one. In some organisations, hiring an agency head is a virtually automatic process: the next school superintendent is simply whoever the assistant superintendent is today. No search consultants or professional vetting is necessary when organisations hire from within. References and recommendations are not so important, if they are used at all. Administrators selected through such internal promotional processes arrive at the agency head position through adherence to preexisting agency norms.

Of course, many agencies sometimes hire from outside and other times promote from within. The important point here is that governments promoting administrators from within tend to apply different selection standards from those recruiting from outside, perhaps, even when comparing internal and external candidates for the same position. The very act of recruiting candidates from outside the organisation indicates, at some level, a demand for innovation.

Adaptation

A simultaneous adaptation process occurs among mobile administrators. Administrators seeking career advancement observe the behaviour of those who successfully "get ahead" and then mimic this winning behaviour. For the mobile administrator, adaptation means building a professional reputation pursuant to higher status jobs since professional credentials and professional reputation are important selection criteria for higher status agency heads. A bureaucrat seeking career advancement via movement to another government will be very active in her profession and will seek to introduce professionally fashionable innovations to her agency. If a profession favours environmental conservation policies, then a reputation for environmental conservation policies is advantageous for the mobile bureaucrat. A bureaucrat hired from outside arrives with a perceived mandate for innovation, for her hiring was due in part to her reputation for professionalism. For administrators not interested in potential movement to a job in another government, adaptation means adhering to organisational norms. A bureaucrat who has advanced vertically within an organisation and who is not seeking job options elsewhere is not so interested in pursuing professionally innovative policies and likely has no specific mandate for innovation.

Conditional Professionalism

Consequently, agencies' policies depend in part on their bureaucratic selection processes and individuals' adaptations to those selection processes. A government that hires its agency heads from outside is most likely choosing to hire a "professional" to provide expert advice and service. An agency head arriving from outside is likely to perceive a mandate for professional innovation and so is more likely to introduce professional innovations than her peers promoted from within. For professions whose policy sensibilities favour environmental conservation, we would expect environmental conservation policies to follow the hiring of an agency head from outside the agency.

The result is mobility-contingent professionalism. Bureaucrats' political decisions must be traceable to their professional identities if professional socialisation is to be useful as a theory of public administration or policy innovation. I argue that professional sensibilities are most likely to materialize as policies where labour market conditions are amenable to them. For political purposes, the mobile administrator is the professional administrator, and professionally sanctioned environmental policies are likely to follow where she goes. With apologies to Forrest Gump, in politics, professional is as professional does.

Conservation Water Rates

For several reasons, water utility service rates provide an excellent subject for evaluation of the theory advanced here. First, water utilities are ubiquitous throughout the United States, and so utility service rates are as well. Although resource and financial conditions vary widely, local governments provide water utility service in virtually every urbanised area in the United States. Wherever water utilities exist, natural resource policies also exist. Few environmental policies are so universally applicable at the local level and, therefore, broadly generalisable in the United States.

Second, though water rate designs are formally adopted into law by elected officials, they are also inescapably technical and are typically developed by and with professional utility managers. Water rates are designed collaboratively, with both administrators and elected officials involved in the process. Berry's (1979) analysis of electricity rates demonstrates that bureaucratic professionalism can significantly affect utility rate design. That administrators are so involved with rate design makes utility rates an excellent place to look for the influence of bureaucratic professionalism in policy making. Utility rate setting is among the most visible, politically sensitive tasks that utility administrators must perform. Although other actors are often involved in the rate design process (e.g., elected officials, lower-level staff members, consultants and citizens), it is difficult to imagine a utility setting its rate structure without the significant involvement of its top administrator.

Third, though American water utilities employ a wide variety of rate structures, they are easily categorised into five basic types:

1. Flat rates, which charge all customers the same amount periodically, regardless of consumption;
2. Uniform rates, which charge a single price for every unit of water consumed at any level of volume;
3. Declining block rates, which charge higher per-unit prices for low volumes of water, but lower per-unit prices at higher volumes;
4. Inclining block rates, which charge progressively higher per-unit prices for water at higher volumes; and
5. Seasonal rates, which charge higher per-unit prices during periods of peak demand or low resource availability, and lower per-unit prices during periods of lower demand or higher resource availability. Seasonal rate variations may be used in conjunction with any of the other four structures.

At the most basic level, principles of price elasticity imply that any rate structure imposing a marginal unit cost greater than zero encourages conservation. However, only the last two structures, inclining block and seasonal rates are considered conservation oriented for purposes of this study because these rate structures are designed with resource conservation in mind. Inclining block rates raise the marginal cost of water consumption at progressively higher volumes, and seasonal rates raise the marginal cost of water during periods of relative resource scarcity. Any rate structure can be designed to generate a desired level of revenue; the choice to increase or reduce rate revenue is distinct from the choice of rate structure.

Like many other environmental policies, public utility rates have important redistributional consequences, for rate designs necessarily affect the allocation of costs and benefits among customers. Utility rates collect revenue as necessary to meet a utility's operating and capital needs. However, utility rates also can be designed to send signals to consumers about socially desirable use of a collective resource. Low prices can promote affordable water for basic sustenance or signal a preference for more water use to encourage development, for example. Higher marginal prices can signal a desire to conserve water. A utility's adoption of a conservation-oriented rate structure is a decision to redistribute resources and so to accept the political risks that accompany redistributive policies. Adjusting water rates in ways that adversely affect high-volume water users is "a political high-wire act" as Postel has observed.

A conservation-oriented water rate structure is exactly the sort of risky, potentially controversial policy that a policy entrepreneur might be expected to champion.

Finally, a distinct and mature water utility management profession exists in the United States, with values and priorities that include water resource conservation. A professional society, the American Water Works Association (AWWA), constitutes and governs the water utility management profession, and it establishes standards of practice, licensure, and ethics that carry the force of law in many states. Dominant norms and values are not static in any profession and, in fact, have changed significantly over time within the water utility profession. Water resource conservation generally, and conservation-oriented water rate structures particularly, have emerged and grown as AWWA priorities over the past three decades. AWWA's general interest conferences routinely feature numerous sessions on conservation, and since 2002 AWWA has sponsored numerous conferences and publications devoted to conservation. The 1972 edition of Manual M1, AWWA's manual of practices on water rates, explicitly discouraged the use of pricing as a means of managing demand or promoting resource conservation (AWWA 1972). By contrast, the most recent edition of Manual M1 devotes six chapters to conservation-oriented rate design, as well as a chapter on securing political support for new rates (AWWA 2000). Conservation-oriented rate structures are among the management practices now clearly favoured by the water utility profession, and this relatively coherent professional norm makes a useful subject of study.

Hypotheses

The following hypotheses relate professionalism and job mobility to the adoption of water conservation rates in American local utilities:

Hypothesis 1—Professionalism: The likelihood of a utility adopting conservation water rates increases as the professional involvement of its agency head increases.

Hypothesis 2—Mobility: Governments that hired agency heads from outside the organisation are more likely to adopt conservation water rate structures than are governments that promoted agency heads from within.

Hypothesis 3—Contingent professionalism: For governments that hired agency heads from outside the organisation, the likelihood of adopting conservation rates increases as the professional involvement of its agency head increases.

The dependent variable for all three of these hypotheses is the adoption of conservation-oriented water rate structures. Hypothesis 1 relates professional identity to the likelihood of conservation water rate adoption. If professional socialisation drives administrators to favour professionally innovative policies, we would expect high levels of professional involvement to be associated with introduction of conservation rates. Hypothesis 2 relates bureaucratic career path or job mobility to the likelihood of conservation water rate adoption. If job mobility encourages the diffusion of professionally fashionable policy innovations, then we would expect governments hiring agency heads from outside to adopt conservation rates more often than those promoting bureaucrats from within. Hypothesis 3 posits a role for professionalism that is contingent upon mobility: the policy effect of an executive's professional involvement depends on his or her career .path. In this way, professionalism becomes manifest in policy when a government signals a demand for innovation by hiring from outside.

An Illustrative Case

Though it sits just 20 miles from the shores of Lake Michigan, the fast-growing Milwaukee suburb of Waukesha, WI (population 67,814), faces a potentially serious water supply shortage. By the late 1990s, long-term draws on the city's aquifer had left the city's groundwater contaminated with potentially unsafe levels of radium. Waukesha's water utility offers an illustration of contingent professionalism at work in conservation policy.

Late in 2002, Waukesha hired Dan Duchniak as general manager for its water utility. At 35, Duchniak was an unusually young hire to head a large municipal utility, but by 2002, he had already served utilities in nearby Racine and Oak Creek for several years. As assistant manager in Oak Creek, he helped develop an innovative storage and recovery system to manage the city's aquifer. Duchniak joined AWWA early in his career and became very active in the organisation, serving as secretary-treasurer of the Wisconsin section and cofounding the Midwest Utility Expo, an annual regional conference for utility operators (AWWA 2007). In 2007, Duchniak ran for and won a seat on the AWWA's national Board of Directors. His record of professional leadership and demonstrated technical proficiency were central to the Waukesha's decision to hire him.

Since taking over the utility, Duchniak has addressed Waukesha's supply shortage by experimenting with new treatment technology and groundwater management techniques and has pursued new sources of supply, including a controversial proposal to draw water from Lake Michigan. More importantly for the present inquiry, under Duchniak's direction, Waukesha has emerged as a leader in water conservation policy. Waukesha recently introduced a rebate program for efficient fixtures and residential irrigation restrictions. Waukesha was not the first or only Wisconsin utility facing supply problems, but in 2007, Waukesha became the first Wisconsin utility to adopt conservation rates, which Duchniak explicitly characterised as a policy meant to help manage demand. "[Waukesha adopted] the most aggressive conservation program in the Midwest," boasted Duchniak. "In the last three years, we cut our [peak demand] by 30 percent and water use by more than 10 percent".

Duchniak's proposal to draw water from Lake Michigan generated opposition from environmentalists and some neighbouring governments. Duchniak's predecessor pursued new supply options too, questioned the US Environmental Protection Agency's (EPA) assessment of Waukesha's water quality, and filed suit against the EPA to challenge its enforcement of the radium contamination standard. But these were not politically risky initiatives since the environmental costs of new supply sources are borne by people outside Waukesha, and lawsuits push environmental issues into a political arena outside local government. By contrast, curbing demand through rates and watering restrictions are politically risky because they impose costs on many of Waukesha's own utility customers. Irrigation restrictions are regulatory policies. Inasmuch as conservation rates impose relatively greater and lower costs on customers with different demand patterns, the new rate structure created relative winners and losers; that is, they are redistributive. Not surprisingly, these policies generated opposition from some community members. Pursuit of these conservation-oriented innovations was politically risky for Duchniak in a way that expanded supply, lawsuits, and technological innovations were not. According to the theory offered here, Duchniak's deep professional involvement disposed him to prefer the professionally sanctioned goal of resource conservation and also gave him ready access to innovations like conservation rates and irrigation restrictions.

Hired from outside the city in the midst of an ongoing supply crisis, Duchniak had an effective mandate for innovation from the elected officials who hired him. In short, the conditions were right for an entrepreneur to emerge in Waukesha to introduce conservation policies to the state of Wisconsin. Duchniak also perceived his aggressive pursuit of conservation as improving his standing in the water utilities profession: at the time of this writing, Duchniak had announced his candidacy for the vice presidency of the AWWA, specifically touting his record on conservation in campaign statements (AWWA 2008).

This case is meant to be illustrative, not demonstrative. That is, Waukesha shows how one professionally active, mobile administrator became an environmental policy entrepreneur. Such illustrative cases "make the unfamiliar familiar" and are chiefly useful in helping to understand other kinds of data. But Waukesha is a single case, and by no means a "hard" one. To gain greater traction on the theory at hand, I turn to data on a wider array of managers and the adoption of conservation rates in their water utilities. Cross-sectional data on four groups of variables are used in this study: (1) water rate structures,

(2) agency heads' professional involvement and career paths, (3) governance structure, and (4) water resources.

Water Rate Structures

Data on rate structures are drawn from an original survey of water utility managers heading American municipal government agencies conducted over the summer of 2006. The survey employed a randomised sample of 150 agency heads, stratified to draw data from agencies of many sizes. A great majority of American water utilities are very small, have little or no professional management, and serve very small proportions of the total US population.

A simple random sample would likely offer relatively little data on large and medium-sized governments due to the very high number of small local governments in the United States. As large and medium-sized utilities serve the majority of the US population, stratifying to ensure their inclusion in the sample is important for drawing broadly generalisable conclusions in policy studies.

Stratification also ensures that data are gathered from agencies occupying every stratum of the water utility profession. The survey was administered via an Internet-based questionnaire. Sampled administrators received a prenotification letter via postal mail approximately 10 days prior to the survey launch and then received an e-mailed invitation to participate in the Internet survey. The response rate was 50.4% of valid cases.

Of the 70 utilities represented in the survey, 22 (31.4%) reported using inclined block rates. Another five utilities reported using a seasonal rate schedule in combination with a rate structure other than inclined block. A total of 27 (38.6%) of surveyed utilities used either inclined block or seasonal rates and so are considered to have adopted conservation rates for the purposes of this study.

These proportions are similar to those generated by AWWA's 2006 Rate Survey, suggesting that the present sample is reasonably representative of the population of utilities, at least with respect to rate structures. At the time of the survey, every participant's current utility rate ordinance or resolution had been adopted during the participant's tenure as head of the local utility. Data on rate structure adoptions cannot show definitively that the administrators were the sole driving forces behind conservation rates. But the involvement of others in rate design does not negate the impacts of executive mobility and professionalism on the process. As noted above, other studies have demonstrated that senior utility managers are critical to the rate setting process. The argument here is that mobility and professionalism make environmental policy innovation more likely, but these forces need not work alone.

Career Path and Professional Involvement

Respondents to the survey were asked about their employment history, including whether they arrived at their current jobs via internal promotion or external recruitment. Sixty percent of respondents reported arriving at their current jobs from outside the agency, with 40% promoted from within. The analyses here use this measurement as the dummy variable outside hire. Professional involvement is measured with an index that compiles the following indicators, as reported by survey respondents.

- The number of professional society memberships that the administrator holds;
- Whether the administrator serves on any committees of the professional society;
- The number of professional conferences attended in the past year;
- Frequency and depth of reading of professional journals; and
- Whether the administrator consulted with professional peers when addressing a policy issue in the past 12 months.

This approach to professionalism is indirect since it uses a behavioural metric (professional involvement) to capture a sociopsychological phenomenon (professional identity). It seems reasonable, however, to assume that those who identify strongly as professionals are relatively active in professional societies and that such involvement reinforces professional identity; certainly, that is the assumption underlying most theories of professionalism resting on socialisation. In any case, this behavioural approach improves upon studies of policy adoption that have used measures of institutional capacity (e.g., agency size, professional recruitment activity) to capture professionalism. Administrators' gender, race, and ethnicity are potentially important independent variables, too. Minority status might affect administrators' opportunities for participation in professional networks and/or the risks of innovation if the political arenas in which they work are systematically biased in against them. Unfortunately, the data used here offer inadequate variation on these variables to analyse their effects since the sample of utility managers is overwhelmingly male and White, non-Hispanic. The very low frequency of women in the sample may indicate that, as Nancy Burns has put it, "gender has already done much of its work" in utility administration careers before respondents ever advanced to a position where they might be surveyed (2002, 467). The same might be said of race and ethnicity.

Governance Structure

Mullin's (2008) recent study argues that specialised local governance structures affect utilities' adoption of inclined block water rates (or "progressive rates," as she calls them). In Mullin's study, the institutional structure of interest is general-purpose municipalities (such as cities, towns, and villages) versus special district governments (such as water authorities and utility districts).

Both general-purpose and special district governments provide water service, but Mullin theorises that special district governments are more likely to adopt progressive rates under conditions of low issue salience because their specialised purpose makes them more attentive to issues (like water rate design) that might fail to draw the attention of general-purpose governments. Differences between special district and general-purpose governments in the probability of adopting progressive rates diminish where issue salience is higher, she argues. The analyses presented here include a dummy variable for special district versus general-purpose governance structure in order to accommodate Mullin's specialised governance theory.

Another institutional feature that might inform the kinds of policies that local governments adopt is council-manager versus mayor-council governance structures. Research on council-manager governments and policy adoption generally focuses on the relationship between elected officials and the city manager, not the effect of institutional form on the behaviour of bureaucratic agency heads. A council-manager structure might reduce the probability of a utility manager introducing conservation water rates. As the professionals with most direct access to elected officials, city managers might restrict department heads' access to the legislative process and reduce policy entrepreneurship opportunities for department heads. Under mayor-council structures, department heads might have more direct access to elected officials and so have greater opportunities for policy entrepreneurship. However, the data analysed here showed no appreciable relationship between conservation rate adoption and council-manager form of government. Moreover, dummies for council-manager form had no significant effect in the multivariate models in this study and so are excluded from the models presented here.

Water Resources

Three variables are included in the models to account for resource scarcity on the likelihood of adopting conservation water rates. Not surprisingly, Hewitt (2000) finds that utilities in hot, dry, sunny climates are more likely to adopt conservation rate structures. Mullin's (2008) analysis of inclined block water rate adoption includes precipitation and daily maximum temperature as variables but uses them to measure the salience of water supply as a political issue, not directly as a need or demand for water conservation.

The present study uses the climatic moisture index ([Im]) developed by Willmott and Feddema (1992) as the main measure of water resource scarcity.

As a metric of resource scarcity, the [Im] has a number of advantages over simple climatic measures like precipitation, temperature, and sunlight because it integrates these variables with the land's water retention capacity and evapotranspiration potential. In this way, the | m "reflects the relationships between climate and the availability of moisture at the earth's surface". Scaled symmetrically about zero and bounded by -1.0 and 1.0, the [Im] also enjoys the advantage of mathematical elegance.

An [Im] value of zero reflects a climate where available water and climatic demand for water are exactly equal. Negative values of [Im] indicate relatively little available moisture, and positive values of [Im] indicate relatively more available moisture.

To put [Im] in more meaningful terms for readers familiar with American geography, the [Im] value is .42 for Seattle, .32 for New York City,. 13 for Omaha, -.08 for Dallas, and -.80 for Phoenix. The mean [Im] in the sample analysed here is .13.

Of course, the [Im] does not account for every potentially relevant climatic condition, and short-term fluctuations in temperature or moisture conditions might cause localised or temporary drought conditions. Neither does the [Im] account for water quality concerns that might drive water scarcity, as they did in the Waukesha case related earlier. However, the [Im] is a scientifically valid metric used by atmospheric scientists and an improvement over simple temperature and precipitation data. The models in this study use the average annual moisture index values for each of the agencies' locations as calculated and published by Willmott and Matsuura in 2007.

The models also include a dummy variable to indicate whether the utility's principal source of water is purchased from a wholesale supplier. Under a wholesale water purchase arrangement, stewardship of the resource rests with the wholesale producer, not the retail utility. To the extent that utilities purchasing water from wholesalers are insulated from resource constraints, they might be less likely to impose water rate structures out of concern for resource conservation. Moreover, many wholesale contracts between utilities employ a fixed payment schedule, so that the payment from the retail utility to the wholesale utility is insensitive to fluctuations in demand. Such contracts remove much of the financial incentive for the retail utility to conserve water. Utilities buying water from wholesale suppliers might be expected to be less likely to adopt conservation rates.

Finally, the effect of utility size is controlled using the natural log of customer connections for each water utility. Larger agencies might be somewhat more likely to adopt conservation rates than their smaller counterparts since design and implementation of conservation rates requires a moderate degree of technical sophistication and capacity. The log transformation is theoretically consistent with the nonlinear nature of agencies' policy needs: differences in agency size should matter less at the high end of the distribution than at the low end. For example, we would expect the substantive difference between a utility with 500 connections and one with 5,000 connections to be greater than the difference between a utility with 30,000 connections and one with 34,500 connections.

Analysis

Since the dependent variable in this study is a binary state (the presence or absence of conservation rates), I test hypotheses using a pair of logistic regression models that predict conservation water rate adoption. Model 1 tests the direct effects of professional identity and career path on conservation rate adoption; Model 2 adds a multiplicative interaction of professional involvement and outside hire.

Professional Identity and Career Path

Hypothesis 1 draws a connection between administrators' professional involvement and water conservation rates. If administrators' professional identities drive policy innovation, then we would expect that higher levels of involvement in the utilities profession to be associated with greater likelihood of conservation rate adoption. The results of Model 1 do not support this hypothesis. In fact, Model 1 finds a slightly negative (albeit statistically dubious) effect of professional involvement on the adoption of conservation rates. This result casts doubt on bureaucratic professional socialisation alone as a driver of agency policy making.

Model 1 affirms Hypothesis 2; however, agencies are much more likely to use conservation rate structures if they hire their agency head from outside the organisation, even when controlling for relevant climatic conditions. With other variables held at their means, Model 1 predicts a .595 likelihood of conservation rates for utilities that hired an agency head from outside, compared to a .059 likelihood for utilities that promoted from within. To relate this finding in more intuitively meaningful terms, present the estimated likelihoods of utilities using conservation rates at different values of the moisture index, with other variables held at their means.

The disparity between utilities with agency heads hired from outside and those with agency heads promoted from within demonstrates the effect of executive job mobility. The differences are greatest and statistical confidence the highest at the middle of the climatic moisture range. At the high and low ends of the moisture scale, the relative abundance or scarcity of water attenuates the differences between agencies with executives hired from within and those with executives hired from outside. But under moderate climate conditions, bureaucratic mobility has a substantial impact on the likelihood of conservation rate-adoption. Agency heads hired from outside are likely to endorse, support, or sustain conservation water rates, even under conditions of relative resource abundance. Coupled with the rejection of Hypothesis 1, this finding suggests that job mobility, not professional identity, encourages the diffusion of environmental conservation policies.

Contingent Professionalism

According to Model 2, professional involvement is a strong, positive predictor of conservation rates for agencies with executives hired from outside and a very strong negative predictor of conservation rates for agencies with executives promoted from within. It depicts the marginal effect of career path on the likelihood of conservation rate adoption generated by Model 2 for bureaucrats hired from outside and for those promoted from within, at varying levels of professional involvement. Career path has little impact on conservation rate adoption for administrators with low professional involvement. The upward slope and narrowing confidence interval (CI) indicates that the substantive and statistical effect of career path increases as professional involvement increases. In other words, a high degree of professional involvement for an agency executive makes more of a difference in policy outcomes when she has been hired from outside the agency.

On its own, professional identity does little to affect conservation rate adoption, as Model 1 shows. For administrators promoted from within, professional values apparently are not reflected in conservation rate design. However, for administrators hired from outside, professional identity is manifested in policy. Also, the effect of job mobility on conservation rate adoption is contingent on professional identity. The positive effect of job mobility on conservation rates in Model 1 turns negative in Model 2. This shift suggests that professional policy innovation does not follow from mobility alone and that mobility in absence of professional identity may actually reduce the probability of conservation rate adoption. Taken together, these results affirm Hypothesis 3 and are powerful evidence for a theory of mobility-contingent bureaucratic professionalism.

Controls

The variables measuring water scarcity generally conform to expectations in both models. The moisture index is a strong predictor of conservation rates in both models, with higher (wetter) values predicting lower likelihood of conservation rates and lower (drier) values predicting higher likelihood of conservation rates. For example, under Model 1, an agency located in a relatively moist place like Allentown, PA, with a moisture index value of 1 standard deviation (SD) above the mean ([Im] = .447), has a. 118 likelihood of adopting conservation rates with all other variables at their means. For an agency in a relatively dry place like Austin, TX, with a moisture index value of 1 SD below the mean ([Im] = -.189), the estimated likelihood of adopting conservation rates is .577. Not surprisingly, at the time of this writing, Austin's water utility used conservation-oriented rates, and Allentown's did not.

Purchased water is likewise associated with lower likelihood of conservation rates in both models. With other variables held at their means, Model 1 estimates a .081 likelihood of conservation rate adoption for utilities that purchase water from wholesale suppliers, compared with .392 for utilities that produce their own water.

These results suggest that wholesale purchases insulate retail utilities from the risk of resource depletion and so reduce the incentive for conservation through rate design. Agency size also is positively correlated with conservation rate adoption in both models. Model 1 estimates a .402 likelihood of conservation rates for a utility serving 25,000 connections, compared with a .544 likelihood for a utility serving 75,000 customers, again with other variables evaluated at their means.

Although the magnitudes of the effects are fairly small, special district governance structure increases the likelihood of conservation rates in both models, though the statistical reliability of the finding is modest. These results are essentially consistent with Mullin's (2008) theory of specialised governance. With their particular focus on water supply, special districts are more likely to adopt potentially controversial innovations like conservation water rates than are general-purpose governments. If professional is as professional does, then job mobility is the mainspring of professionalism. Scholars of professionalism have observed that a profession's economic viability and political power rest on an ability to develop a clientele (or market) and define problems in scientific terms demanding technical solutions. Several studies of bureaucratic politics argue that professionalism and professional identity explain the policy choices that administrators make. But this study's findings suggest that professionalism is unlikely to have much impact on the diffusion of politically risky innovations to governments unless the professionals enjoy mobility from one agency to another.

Hiring an administrator from outside involves an assessment of professional reputation, and so an external hire signals, explicitly or implicitly, a desire for the kind of policy innovation for which professions are known. Professional sensibilities may abound among bureaucrats everywhere, but job mobility offers license.to put professional principles into practice and so opens the door for policy entrepreneurship. Clearly bureaucratic professionalism, job mobility, and policy entrepreneurship are parts of a much more complex puzzle. But when a government hires a professional from outside, professionals are likely to take entrepreneurial political risks, and so the priorities of profession are more likely to become manifest as policy.

These results have significant implications for the diffusion of environmental policy innovation. Entrepreneurs figure importantly in accounts of environmental policy development, in part because such policies are usually regulatory and/or redistributive and, therefore, are politically costly. Bureaucratic professions that embrace environmental conservation policies may facilitate the diffusion of environmental policies across governments. But professionalism alone is unlikely to spur environmental policy innovations, at least when they are politically risky. Professions are most likely to gain traction in the environmental policy arena when job mobility affirms and provides incentives for bureaucratic policy entrepreneurs.

Limitations and Outstanding Questions

Important questions about bureaucratic mobility, professionalism, and policy innovation remain outstanding. This study demonstrates a link between public administration career paths and the diffusion of fashionable conservation policy innovation in the water utilities profession. Similar studies could investigate whether the same patterns hold for other professions (e.g., public health, schools, firefighting), as well. As a profession, water utility management is relatively consistent in its norms, and its predominant policy sensibilities are codified in accessible, formal publications. Moreover, AWWA standards and licensure carry the force of law in many jurisdictions. But such coherence and clout are not uniform across professions and, in fact, may be important variables. What does it mean to advance a professionally sanctioned innovation in a profession where norms are fluid and positions on important issues are unsettled or highly controversial within the profession? Studies of diffusion and mobility across multiple professions could offer traction on this question.

Also left unclear are the implications of mobility-contingent professionalism for governments that feature little or no bureaucratic job mobility. Intergovernmental job mobility is highly constrained or nonexistent in many state and federal government bureaus, especially in military, intelligence, and law enforcement agencies. If professional innovation is to a significant degree contingent upon job mobility, then for better or worse, innovation and professionalism may be severely blunted in agencies with little or no mobility. The extent to which this theory is applicable or testable in such agencies is unclear.

Finally, the cross-sectional data used in this study limits the analysis to an environmental policy that was professionally fashionable at the time of the survey. The limitations of cross-sectional research design on studies of policy diffusion are well understood. This study's use of cross-section data avoids some common pitfalls, however, because the rates in effect at the time that the data were collected were adopted during the tenure of the surveyed bureaucrats, and we may reasonably assume that the utility managers were involved in the rate-setting process. Nonetheless, a study of conservation rate structure adoption over time that traced the career paths and policy decisions of specific bureaucrats would further enhance our understanding of bureaucratic professions and policy diffusion.

As long as governments have tried to fashion themselves as democracies, theorists have recognised a tension between efficiency and accountability. With the emergence of public administration professions, theorists have recast this tension as between bureaucratic professionalism and democratic responsiveness. The enduring reality that professional bureaucrats are deeply and pervasively involved in initiating policy is a potential threat to the democratic responsiveness that elections and legislatures are meant to provide. Nonetheless, bureaucratic professionalism is hard to dismiss as hostile to good governance, as Wirt (1985) observes: The professional executive officer is the invisible actor in urban decision making, recalling Shelley: "the awful shadow of some unseen Power floats, though unseen, among us." The professional's "shadow" may be "awful" to those who see in it insensitivity and obstruction of local values and control. But it is difficult to escape the impression that many benefits arrived on the local scene in the briefcases and journals of the professional executive officer.

The present study's contribution to the professionalism-versus-responsiveness debate is the finding that professional bureaucrats are not simply loyal to their professional norms but rather are responsive to their professions when employment conditions give them political license to innovate. Mobility-contingency thus softens the conflict between professionalism and responsiveness, and across thousands of local governments in the United States, perhaps, balances these oft-conflicting values.

Clean Water Act Section 401 and Nonpoint Source Pollution

The Clean Water Act (CWA) was passed to "restore and maintain the chemical, physical, and biological integrity of the Nation's waters." Few Americans would disagree that this was a laudable goal, and America's waters have improved significantly since the enactment of the CWA in 1972. However, the focus of the CWA is point source pollution, and many of the nation's waters remain impaired because of nonpoint source pollution. Unlike point source pollution, nonpoint source pollution cannot be traced to a specific source and is often described in terms of diffuse runoff. According to the General Accounting Office, the leading cause of nonpoint source pollution is agriculture. In fact, a 1995 General Accounting Office briefing report indicated that agricultural nonpoint source pollution impaired seventy-two percent of the miles of affected rivers and streams, fifty-six percent of the affected lake acres, and forty-three percent of the square miles of affected estuaries. Obviously, something must be done about nonpoint source pollution.

Environmentalists advocate using section 401 of the CWA to address the nonpoint source pollution problem. Section 401 requires anyone applying for a federal permit or license for any activity that may result in a discharge to obtain certification from the state in which the discharge originates. Traditionally, section 401 applied only to point source pollution. Some commentators, however, interpret the Supreme Court's treatment of section 401 in PUD No. 1 of Jefferson County v. Washington Department of Ecology (PUD No. 1) as blessing the use of section 401 for nonpoint source pollution. Proponents of applying section 401 to nonpoint source pollution also point to Oregon Natural Desert Ass'n v. Thomas (ONDA I), in which a federal district court held that section 401 applies to grazing permits on public lands. The Ninth Circuit Court of Appeals declined to adopt that interpretation in Oregon Natural Desert Ass'n v. Dombeck (ONDA II). However, that opinion was withdrawn in response to the plaintiffs petition for reconsideration.

Although the holding in ONDA H persuasively determined that section 401 does not apply to nonpoint source pollution, the question remains open pending reconsideration. Another unresolved question is whether the application of section 401 would have any effect on agricultural nonpoint source pollution. Targeting agricultural users of federal lands with more unnecessary government regulation will do little to reduce nonpoint source pollution. Time, money, and energy needed to deal with nonpoint source pollution on farm and ranch land is wasted fighting court battles over the applicability of section 401, and other ways must be found to deal with the problem.

Traditional Applications of Section 401

The full text of section 401 requires that:

Any applicant for a Federal license or permit to conduct any activity including, but not limited to, the construction or operation of facilities, which may result in any discharge into the navigable waters, shall provide the licensing or permitting agency a certification from the State in which the discharge originates or will originate, or, if appropriate, from the interstate water pollution control agency having jurisdiction over the navigable waters at the point where the discharge originates or will originate, that any such discharge will comply with the applicable provisions of section 301, 302, 303, 306, and 307 of this title.

Section 401 ensures that federal agencies cannot override the states' water quality standards. It allows states to certify the activity first, to ensure that the activity meets the specific provisions of the CWA and the states' water quality standards. This allows for a balance between the states and the federal government, similar to the balance preserved in the section 402 permit process of the CWA. Section 402 allows either the Environmental Protection Agency

(EPA) or a state, if it has been approved to do so, to issue permits for the discharge of pollutants. The national pollution discharge elimination system (NPDES) in section 402 was designed to prevent polluters from "shopping around" for a state with a weak permit program by allowing EPA to withdraw a state's permitting authority if the state was not running its program correctly. In a similar manner, section 401 maintains a balance between the states and the federal government.

If a state has a weak program, the federal government still reviews the activity after the state has issued its certification. Likewise, if the federal permitting agency is not taking water quality standards seriously, the state can impose conditions under section 401 to prevent degradation of its waters. Technically, section 401 applicants must apply for state certification first. Once state certification is received, the applicant sends it with a federal permit or license application to the responsible agency. If the state denies certification, the project is essentially dead.

In reality, the section 401 process does not always work this way. States sometimes allow responsible federal agencies to review an application first. Following federal review, the application goes to the state for, in essence, rubber-stamp certification. States reason that federal agencies have the expertise to determine whether a project meets the provisions of the CWA. While this may not be how Congress intended states to use section 401, it still allows states to maintain some control over the quality of their navigable waters.

Sections of the CWA that States Must Consider in Granting a Section 401 Permit

Section 401(a) outlines certain provisions of the CWA that states must consider in the certification process. These provisions deal almost exclusively with point source pollution. Section 301, the first listed provision, makes the discharge of pollutants unlawful and sets up a two phase program for applying effluent limitations. "Any restriction established by a State or the Administrator [of the EPA] on quantities, rates, and concentrations of chemical, physical, biological, and other constituents which are discharged from point sources into navigable waters" is an effluent limitation. Obviously, section 301 deals with regulation of point source pollution. The same is true of section 302, which allows for stricter effluent limitations if the "discharges of pollutants from a point source or group of point sources" would affect a certain portion of a navigable waterway and would interfere with the designated uses of that waterway. Section 306 establishes federal standards of performance for controlling discharge of pollutants. The CWA defines the phrase "discharge of pollutants" as "any addition of any pollutant to navigable waters from any point source." Therefore, this section also applies to point source pollution only. Section 307 establishes toxic and pretreatment effluent standards. This section subjects each toxic pollutant to effluent limitations, which, as previously noted, are limitations on discharges from point sources. Pretreatment standards are established to prevent the discharge of pollutants from treatment works, and treatment works are point sources under the CWA. Therefore, section 307 deals with point sources as well.

Lastly, section 303 requires the states to establish water quality standards and to implement plans to achieve these standards. While effluent-limitation-based regulation controls water pollution by regulating the amount of pollutants discharged by a particular source, water quality standards control pollution by regulating the amount of pollutants discharged into a particular segment of water. Therefore, water quality standards could include point and nonpoint source pollution. However, section 303 only mentions effluent limitations, which apply to point source pollution. Specifically, the section requires each state to look at its own waters and determine if the effluent limitations required by section 301 are strict enough to meet state water quality standards.

The provisions of the CWA that a state must look at before issuing a section 401 certification basically address point source pollution. Because of this, when the states use their section 401 power, they apply it to activities that result in discharges from point sources and not to activities that result in nonpoint source pollution.

The Three Traditional Uses of Section 401

States use their section 401 power in three federal permit and license processes. These are section 402 permits issued by EPA, section 404 permits issued by the Army Corps of Engineers (Army Corps), and hydroelectric plant permits issued by FERC. All three permits deal with point source pollution.

Section 402 Permits

National Pollution Discharge Elimination System (NPDES) permits issued under section 402 allow the discharge of pollutants as long as the discharge meets established effluent limitation standards. Section 402 permits are solely for discharges from point sources, and most states are authorised to issue them. However, EPA does issue some section 402 permits, and by regulation the state of origin must issue a section 401 certification first. In fact, if EPA receives a request for a 402 permit, EPA's Regional Administrator is required to send the request to the state for certification. In essence, EPA forces the states to use section 401.

Section 404 Permits

Another area in which states have traditionally used their section 401 power is with section 404 permits. The Army Corps of Engineers is authorised to issue or deny section 404 permits for the discharge of dredged or fill material into navigable waters at specified disposal sites. These are point source discharges because a discernible conveyance is used. Section 401 puts section 404 within its scope when it discusses "use of spoil disposal areas." In addition, Congress made it abundantly clear during the legislative work on the CWA in 1972 that section 401 was to cover section 404 permits. A proposed amendment would have exempted dredging from the section 401 process, but the amendment never became part of the Senate bill, and the states retained authority to certify section 404 permits under section 401.

FERC Permits

The last traditional use of section 401 is for FERC permits for hydroelectric projects. The controversy surrounding hydroelectric projects centred around dams. In National Wildlife Federation v. Gorsuch, EPA maintained that dams fell under their definition of nonpoint source pollution. Environmentalists disagreed with this interpretation and argued that a dam was a point source. The D.C. Circuit held that the statute would support defining hydroelectric projects as either a point or nonpoint source.

The court deferred to the agency's position because it was not unreasonable. Although Gorsuch dealt with a section 402 permit, not section 401 certification, the question the court was deciding was whether a dam was a point source. The court deferred to the agency's interpretation that a dam was a nonpoint source. However, most people would probably agree with the environmentalists' interpretation that a dam is a point source.

A dam falls under the definition of a point source because it is a single structure that has observable effects on navigable water, and those effects can be modified through technology. Certainly, the building and operation of a hydroelectric project result in point source discharges. Although hydroelectric projects may result in some nonpoint source pollution, taken as a whole hydroelectric projects are point sources.

Ultimately, EPA backed away from its original stance that a hydroelectric project is not a point source. Accordingly, everyone seems to be in agreement that a dam is a point source. Therefore, this traditional application of section 401 permits also relates primarily to point sources only.

The States' Power Under Section 401

The use of section 401 is left up to the states. Each state may decide whether to waive certification for an activity. Sometimes, waivers result simply from a state's inaction. If the state does not act on the request for certification within a year, the federal permitting agency will consider certification waived. Although states use section 401 in different manners, most seem to have confined themselves to the three traditional uses of section 401 mentioned above.

The Emerging Effort to Apply Section 401 to Nonpoint Source Pollution

PUD No. 1 and ONDA I represent a growing interest in applying section 401 to nonpoint source pollution. As the previous section discussed, states have been applying section 401 only to point source discharges. However, many say that states can, and should, apply section 401 to nonpoint source pollution, arguing that PUD No. I supports that position. The following discussion looks at the holdings in PUD No. 1 and ONDA I in connection with section 401 and nonpoint source pollution.

PUD No. 1

The PUD No. i case addressed the section 401 permit process for a hydroelectric project on the Dosewallips River. The Washington Department of Ecology, the state agency responsible for section 401, conditioned the permit on minimum stream flows. At issue in the case was whether the Washington Department of Ecology could condition a section 401 certification on the maintenance of minimum stream flows to protect fish habitats.

The Court's Holding

PUD No. I was not about whether a section 401 certification was needed for the project. Washington state was using its section 401 power in a traditional manner to certify a hydroelectric project prior to FERC granting a permit. Although the need for a section 401 certification was not in question, the Court did touch briefly on this issue. The Court used the traditional two-part analysis to determine if a section 401 permit could be required by the state. This analysis is relatively simple. A court must first determine if a federal permit or license is necessary for the project. Then it must determine if some kind of discharge may result from the project. Here, the Court found that the project would need a FERC license and that at least two discharges would result from the project, so Washington could require a section 401 certification.

The question for the Court was not whether a section 401 certification was required, but exactly what the state could consider in granting it. In PUD No. I the state wanted to impose a minimum stream flow between one hundred and two hundred cubic feet per second (cfs). The Court allowed the condition under section 401, recognising that a state may condition its permit not only on the specific provisions of the CWA outlined in section 401(a), but also on "any effluent limitations and other limitations" needed to meet state water quality standards. The Court interpreted this provision of section 401 to mean that the state could consider the effect that the entire project or activity would have on water quality, not just effects from specific discharges. But, in order for the state to consider the effects of the activity as a whole, the activity must first be one that would fall under section 401, as the building of PUD No. 1 did in this case. This interpretation by the Court to allow the state to look at the effects of the activity as a whole does not expand the scope of section 401 to include nonpoint source pollution. The court's interpretation only expanded what states can consider when considering point source certification.

The Court also addressed whether the state was preempted by the Federal Power Act because FERC has comprehensive authority to issue permits for hydroelectric projects and is required to look at the effects the project will have on wildlife and fish habitats.

The Court held that no preemption problem existed because FERC had not yet acted on the project license. In the end, the Court's holding concerned the scope of a state's power under section 401. The holding did not address what kind of discharge, point source or nonpoint source pollution, triggers the section 401 permit requirement.

The Holding as Applied to Nonpoint Source Pollution

Commentators who advocate expanding the use of section 401 are not interested in the PUD No. I holding. Rather, they are interested in the Court's brief mention of the two discharges that would result from dam construction: "the release of dredged and fill material during the construction of the project and the discharge of water at the end of the tailrace after the water has been used to generate electricity."

Some commentators believe the second of these two discharges can be interpreted to include nonpoint source pollution, even though the case did not hold that and the Court never specifically addressed point source versus nonpoint source pollution.

Discharge from a Tailrace Is Not Nonpoint Source Pollution

Water coming out of a tailrace has a changed temperature, and this can affect the ability of the water to support certain fish. A tailrace is defined as "a race for conveying water away from a point of industrial application (as a waterwheel or turbine) after use."

A race is defined as "a strong or rapid current of water flowing through a narrow channel." The narrow channel created to carry the water away from a hydroelectric project is normally a tunnel, some kind of pipe, or a ditch. Each of these items is a "discernible, confined and discrete conveyance" within the definition of a point source in the CWA. Therefore, a tailrace is a point source. In fact, in one PUD No. I brief, tailrace is defined as a conveyance for "water that has passed through the turbine." And if a tailrace is a point source, then the water coming out of it is discharged from a point source, and is not a nonpoint source. So, the Court's mention of discharge from the tailrace does not indicate an expansion of section 401 to nonpoint source pollution.

The Power Authority of New York Case: Tailrace Water Is Not Nonpoint Source Pollution

At least one commentator asserts that most courts would hold that water coming out of a tailrace is nonpoint source pollution because water temperature has been changed and temperature is not an identifiable pollutant. This assertion is based on Power Authority of New York v. Williams in which the court addressed the denial of a section 401 permit for a hydroelectric project. One of the reasons for the denial was the Power Authority of New York's failure to assure the state that water temperatures in the affected waters would be kept at a satisfactory level.

The Power Authority of New York argued that water temperatures could not be considered under state water quality standards because a section 401 permit required that the discharge involved contain an identifiable pollutant. The court rejected this argument holding that "a transfer of water from the upper reservoir to the lower reservoir is a `discharge'" because the water transfer fell within the state's definition of industrial waste. Industrial waste includes any liquid from any process of industry that may cause water pollution, and under the state's section 401 regulations, a permit is required to discharge industrial waste. Thus, the court held that the transfer of the water from one reservoir to another could be a discharge that would trigger a section 401 permit even though it contained no specific pollutant.

Without ever mentioning point sources or nonpoint sources, the court's reasoning supposedly supports the contention that section 401 applies to a discharge regardless of its source. However, this contention makes little sense because the water was transferred by a pipe, which is a point source, and a discharge from a point source must contain a pollutant to be covered under the CWA.

So, although the court held that this discharge contained no specific pollutant, it defined the discharge as an industrial waste. Industrial waste is one of the specific pollutants listed in the definition of pollutant in the CWA. Thus, water with a lowered temperature is a point source discharge, not nonpoint source pollution. Therefore, Power Authority of New York does not support the contention that water from a tailrace would be viewed as nonpoint source pollution.

PUD No. I never specifically addressed whether section 401 applies to point sources or nonpoint source pollution. All the Court did was mention two types of discharges from this specific project and give some examples of other statutes that set up federal permits which may require 401 certification. Trying to interpret its holding as a blessing of the use of section 401 for nonpoint source pollution is a stretch indeed. All PUD No. I did was increase the scope of section 401 certification conditions.

Caswell Shows PUD No. I Did Not Decide Section 401 Applies to Nonpoint Source Pollution

In Idaho Conservation League v. Caswell, a group of environmentalists who tried to use PUD No. I to support a claim that section 401 applies to nonpoint source pollution lost their case on summary judgment. The environmentalists sued the United States Forest Service (Forest Service) because it did not require a timber company to obtain section 401 certification prior to building two logging roads on Forest Service land. The discharge in question was nonpoint source pollution. First, the court looked at the language of section 401 and found that it was "evident that section 401 was only intended to encompass those projects which resulted in a `point source discharge.'" Then, the court dismissed the environmentalists' reliance on PUD No. I because it "does not stand for the proposition that all nonpoint sources are subject to section 401 certification." This is supported by the ONDA I court, which did not rely on PUD No. 1 to decide that section 401 does apply to nonpoint source pollution.

ONDA II: The Correct Interpretation of Section 401

In section 401, the nonpoint source pollution dispute centres on the meaning of the word "discharge." In concluding that discharge in section 401 does not apply to releases from nonpoint sources, the Ninth Circuit examined "the language of the governing statute, guided not by a single sentence or member of a sentence, but by look[ing] to the provisions of the whole law, and to its object and policy." Rather than focusing on the definition of discharge, as the lower court had done, the Ninth Circuit focused on the intent and purposes of the CWA as a whole. ONDA H does not go far enough because both the definition of discharge and related terms, and the legislative history of section 401 mandate the same conclusion.

The Ninth Circuit Decision

In some respects, the Ninth Circuit's analysis was as cursory as the lower court's. The court emphasized that the CWA "banned only discharges from point sources" and did not directly regulate nonpoint source pollution. The court pointed out that another provision of the CWA, section 208, was designed to deal with nonpoint source pollution. The court also drew support for its conclusion from the fact that in 1987, section 1329 was added to the CWA to require states to adopt nonpoint source pollution management programs. This consistently separate treatment of point and nonpoint sources in the CWA led the court to conclude that the term "discharge" does not include nonpoint sources. The court rejected the environmentalists' argument that nonpoint sources were somehow swept into the scope of section 401 by the section's reference to section 303. The court noted that before 1972 this provision required a "state to certify that a licensed activity would not violate applicable water quality standards." In 1972 the provision was changed to require a state to certify that a licensed activity complies with the applicable provisions of sections 301, 302, 303,306, and 307. The change was made "to assure consistency with the bill's changed emphasis from water quality standards to effluent limitations based on the elimination of any discharge of pollutants." The change, according to the court, was not made to bring nonpoint sources into the scope of section 401, especially in light of the Ninth Circuit's holding in Oregon Natural Resources Council v. United States Forest Service. In that case, the Ninth Circuit held that reference to water quality standards in section 301 "did not sweep nonpoint sources into the scope of [sections] 1311." Similarly, the court held that the mention of water quality standards in section 303 did not include nonpoint source pollution. Thus, all the sections cross-referenced in section 401 deal with point source regulation, not nonpoint source pollution.

Like the Caswell court, the Ninth Circuit found any reliance by the environmentalists on PUD No. i misplaced. Both discharges mentioned by the Supreme Court, according to the Ninth Circuit, were discharges from point sources. Thus "the Supreme Court in PUD No. I did not broaden the meaning of the term `discharge' under [section 401]." The Ninth Circuit also found support for its conclusion in the CWA's use of the term "runoff" as compared to the term "discharge". According to the Ninth Circuit, when Congress meant point source pollution it used discharge, and when it meant nonpoint source pollution it used runoff. Therefore, if Congress had intended section 401 to include nonpoint sources as well as point sources, it would have used language similar to section 313. In that provision, federal agencies "engaged in any activity which may result in the discharge or runoff of pollutants" are required to comply with applicable water quality standards. Had section 401 used such language, then it would include nonpoint sources. But use of the term discharge alone limits its application to point sources.

Lastly, the Ninth Circuit rejected the plaintiffs' argument that discharge is defined more broadly than the discharge of pollutants from point sources. The court flatly stated that this view "is incorrect." According to the court, the broader meaning of discharge concerns whether releases from point sources are polluting or nonpolluting. Discharge includes both polluting and nonpolluting releases from point sources but cannot be read to include nonpoint sources. Any other interpretation would not "comport with the structure and lexicon of the Clean Water Act." By examining section 401 in light of the CWA as a whole, the Ninth Circuit concluded that section 401 does not apply to nonpoint source pollution. While some may fault the court's reasoning and analysis, a closer look at the statutory language of section 401 and the legislative history relating specifically to section 401 shows that the Ninth Circuit was correct.

A Closer Look at Section 401: The Definition of Discharge

"The term `discharge' when used without qualification includes a discharge of a pollutant, and a discharge of pollutants." That is all the definition section of the CWA says about discharge. Yet, the meaning of discharge is important to understanding section 401 because the section does not mention the source of the discharge, only the discharge itself. As the Ninth Circuit noted, the meaning of discharge literally holds the key to understanding the type of source to which Congress was referring in section 401.

The Water Quality Improvement Act of 1970, which first established what is now section 401, did not define discharge. However, in 1972 a definition for discharge was added. How that definition came about sheds considerable light on what it means. The Committee on Public Works in the Senate wrote that the definition of discharge was added to define the scope of the control that would be required under the CWA. It then discussed pollutants added by any point source, outfall or other pipeline, and publicly owned treatment works. Discharge was said to be a term of art in the legislation that "refers to the actual discharge from a point source into the navigable waters, territorial seas, or the oceans." The report seems to limit the meaning of discharge to point sources only.

The actual definition includes two other terms that are defined in the CWA, "discharge of a pollutant" and "discharge of pollutants." These two terms are defined to mean "(A) any addition of any pollutant to navigable waters from any point source, (B) any addition of any pollutant to the waters of the contiguous zone or the ocean from any point source other than a vessel or other floating craft." Both of these terms limit their meaning to point sources only.

These definitions, along with the legislative history concerning the actual definition of discharge, would seem to end the inquiry as to whether or not discharge applies to nonpoint source pollution. However, the definition of discharge "includes" the discharge of a pollutant and the discharge of pollutants. And, "the use of the word `includes' rather than `means' in a definition indicates that what follows is a nonexclusive list which may be enlarged upon." So, the term "discharge" is not limited to the meaning given to "discharge of pollutant" and "discharge of pollutants."

Environmentalists argue that discharge was not meant to be limited to point sources. They point to the fact that the definition of discharge in the CWA is the only one that uses the term "includes." All the other definitions use the term "means," which is a limiting word in definitions. Therefore, Congress must have intended discharge to include, but not be limited to, point sources. Otherwise, it would have used the term "means" in the definition of discharge. While the absence of nonpoint source pollution from this list is not conclusive evidence that it was meant to be excluded, it does not mean that nonpoint source pollution was included in the definition of discharge. After all, noscitur a sociis, a traditional canon of statutory construction, provides that words in a list should be interpreted in light of the surrounding words. The words in the list of what is included in the definition of discharge all deal with point sources. To imply that nonpoint source pollution would also fit in that list violates this basic rule of statutory construction.

Discharge is further limited by the use of the word "pollutant" in its definition. In the CWA, pollutant "means dredged spoil, solid waste, incinerator residue, sewage, garbage, sewage sludge, munitions, chemical wastes, biological materials, radioactive materials, heat, wrecked or discarded equipment, rock, sand, cellar dirt and industrial, municipal, and agricultural waste discharged into water." Because this definition uses the limiting term "means," only substances that appear on the list are pollutants under the CWA. The listed pollutants came from the Refuse Act, with the exception of "municipal discharges," which was added to the definition in 1972. The Refuse Act requires permits for industrial discharges of refuse into navigable waters.

These industrial discharges come from point sources. The same would be true of any municipal wastes that would be discharged. Therefore, the definition of pollutant only encompasses discharges from point sources. This further limits discharge because its definition only mentions discharging pollutants. If, per the definition of pollutant, a pollutant can only come from a point source, then discharge can only apply to point sources, not nonpoint sources.

In addition to the Ninth Circuit's ONDA II decision, other case law supports a limited meaning of discharge. Commonwealth of Pennsylvania v. Harrisburg held that the plain meaning of the statute meant that a discharge was a "discharge of pollutants" which is defined to include a point source only. Idaho Conservation League v. Caswell also held that discharge was limited to point sources only. Lastly, National Wildlife Federation v. Gorsuch hesitated to add new meaning to the definition of pollutant, which would by inference expand the meaning of discharge. Therefore, based on the plain meaning, legislative history, and case law, the definition of discharge is limited to discharges from point sources only.

Legislative History of Section 401

The statutory language and corresponding legislative history of section 401 indicate that section 401 was not meant to include nonpoint source pollution. However, even if the language itself was not clear, the legislative history behind section 401 further supports the contention that the section was not meant to be used to regulate nonpoint sources.

1970 Amendments

As previously mentioned, section 21(b) of the Water Quality Improvement Act of 1970 was the forerunner for section 401. Section 21 was designed to encourage cooperation by all federal agencies in the control of pollution. The House report mentioned a "wide variety of licenses and permits" issued by federal agencies that may affect water quality. While the legislative history did not exclude permits for nonpoint source pollution activities, the only permits specifically mentioned were for industrial and power plant sources that cause thermal pollution. Obviously, an industrial plant or a power plant is a point source, not a nonpoint source.

The concern was that the permit process between state and federal agencies for permits issued by the Atomic Energy Commission would be duplicative. Section 21 was not meant to be merely a duplication of another regulation. Congress was only concerned with trying to achieve "reasonable assurance ... that no license or permit will be issued by a Federal agency for any activity that through inadequate planning or otherwise could in fact become a source of pollution." Referring to "sources of pollution" also indicates that in section 21, the legislature was concerned with what would become known as point source pollution, not nonpoint source pollution.

Because the term "point source" was not used in 1970, environmentalists contend that section 21(b) was not limited to discharges from point sources. Because Congress did not differentiate between point and nonpoint sources in 1970, it could not have intended section 21(b) only to regulate discharges from point sources. This argument ignores the thrust of section 21(b) and the 1970 amendments in general. In enacting section 21(b), Congress was concerned about projects that the Atomic Energy Commission was building. While Congress emphasized that section 21(b) was not limited to just those projects but to any federal project, nothing indicated that it was also thinking about nonpoint source pollution, especially because nonpoint source pollution was not the focus of the 1970 Amendments.

1972 Clean Water Act

In 1972, Congress revamped the water pollution control statutes, resulting in the passage of what are now known as the Clean Water Act (CWA). Basically, section 21(b) of the Water Quality Improvement Act of 1970 became section 401 of the CWA. The Senate modified the language to ensure consistency with the bill's emphasis on effluent limitations as opposed to water quality standards. The CWA defines the term "effluent limitations" as restrictions on discharges from point sources only. So, the Senate only appeared to be considering point sources when it reworked section 401. The debate in the House centred on whether section 401 was to include discharges from point sources necessary to meet the provisions of sections 306 or 307, both of which deal with point sources. The Senate version of the bill did not include compliance with sections 306 and 307. However, the conference report accepted the House version, which included compliance with two more point sources. Over and over again, the emphasis in section 401 was on point source pollution.

In fact, in 1972 the concern was controlling point source pollution: "the heart of the bill is the new focus on point source limitations." During the debate on the amendments, the fact that the bill did not cover nonpoint source pollution and was not meant to cover nonpoint source pollution came up again and again. That Congress actually meant to include nonpoint source pollution in section 401 when it incorporated section 21 (b) from the 1970 bill into the CWA seems hard to believe, especially because the focus of the bill was not nonpoint source pollution.

Environmentalists argue that because Congress was aware of the problems of nonpoint source pollution in 1972, it surely meant to include nonpoint source pollution in section 401. They argue that this interpretation fits into the general statutory scheme that Congress developed in 1972 for nonpoint source pollution. Congress established section 208 of the CWA to deal with nonpoint source pollution. Section 208 required that the states create areawide waste treatment management plans that include nonpoint source pollution. In essence, the control of nonpoint source pollution was given to the states. Environmentalists argue that Congress would not limit the states then to reviewing point sources when they reviewed federal activity under section 401.

Congress specifically gave the states the authority and power to control nonpoint source pollution through section 208. If the control of nonpoint source pollution was already part of the CWA, then section 208 would be unnecessary. The fact that Congress enacted section 208 supports the contention that section 401 was not meant to include nonpoint source pollution. Because other sections of the CWA did not deal with nonpoint source pollution, Congress had to create a specific section to deal with the problem. Therefore, in 1972 Congress did not intend nonpoint source pollution to be included in section 401.

1977 Amendments

The 1977 amendments to the CWA reinforce the interpretation that Congress did not intend section 401 to include nonpoint source pollution. The only change made to section 401 itself was the addition of section 303 to the list of other provisions of the CWA that a state must consider when granting a section 401 certification. Section 303 deals with water quality standards. Congress originally intended water quality standards to be considered during the section 401 process; however, the states were confused about Congress's intent. To end the confusion, Congress added section 303 to the list of provisions in section 401. Congress felt that it was reasonable to have states include water quality standards when looking at the permit process. This did nothing, however, to increase the scope of section 401 to include nonpoint source pollution. It simply meant that when looking at a section 401 certification, a state should look at its section 303 water quality plan to make sure the activity complies with that plan. This change in section 401 was for clarification, not expansion.

The 1977 amendments did, however, address nonpoint source pollution, just not in the debate surrounding section 401. Agricultural nonpoint source pollution, in particular, was recognised as a major problem in the fight for clean waters. In fact, EPA testimony during the hearings on the 1977 amendments centred around this problem and called for cooperation among federal agencies to find ways to address the problem. However, in its testimony, EPA stressed that section 208 of the CWA, not section 401, would help solve the problem. So, in 1977 Congress was concerned about nonpoint source pollution but chose to address it through section 208, with EPA's backing.

1987 Amendments

The 1987 amendments to the CWA continued to focus on nonpoint source pollution, but once again, not through the use of section 401. Again, Congress recognised that nonpoint source pollution was a problem in need of a solution. Therefore, in the 1987 amendments, Congress created section 319 of the CWA. Section 319 established a national policy and programs for the control of nonpoint source pollution through the creation of nonpoint source pollution management programs. If Congress had intended section 401 to address the problem of nonpoint source pollution, then it could have turned to that section as part of the solution. However, in 1987 Congress never even looked at section 401 as part of the solution.

The legislative history indicates that Congress never intended section 401 to apply to nonpoint source pollution. Since the section's inception in 1970, Congress has considered section 401 to be part of the control mechanisms for point source pollution only. However, there is no reason to believe that environmentalists will not continue their efforts to apply section 401 to nonpoint source pollution, especially if the Ninth Circuit changes the ONDA II holding on reconsideration. The following discussion looks at the implications of applying section 401 to nonpoint source pollution, focusing particularly on the implications to the agricultural community.

2

Assessment of Drinking-Water Supplies

The U.S. Department of Energy's (DOE's) Hanford Site was established in southeastern Washington State in 1943 to house facilities for producing and processing nuclear materials for the U.S. government. Nine plutonium production reactors were constructed on the site between 1944 and 1954, along with facilities to separate and purify reactor products into desirable forms. Eight of the nine production reactors were shut down by the early 1970s. One remained operational until 1987. Decades-long nuclear-fuel processing activities on the site generated an estimated 1.4 billion cubic metres of chemical and radioactive waste in liquid, gaseous, and solid forms. These materials were released to the air, soil column, and surface water, or stored in 177 underground tanks with capacities ranging up to 3.8 million litres. The current DOE mission at Hanford emphasizes cleaning up the tank, the former waste disposal locations, and the groundwater underlying the site, with the intent to limit the movement of persistent contaminants from the soil column (vadose zone) to groundwater and to reduce the transfer of groundwater contaminants to the adjacent Hanford Reach of the Columbia River. Numerous groundwater discharges are known to enter the Columbia River along the Hanford Site shoreline, and Hanford-produced contaminants are measurable in river water near and downstream from the site.

Hanford is an EPA Region 10 Superfund site with a cleanup budget that was approximately 1.9 billion dollars in fiscal year 2003. Activities associated with this classification have, in recent years, attracted many new workers to the site. Anticipated non-DOE uses of cleaned-up Hanford lands may bring in even more workers in future years. Because of the large amount of nuclear and chemical waste still present at Hanford, there is concern about potential health impacts to workers and the public from exposure to remaining contaminants. One area of concern has been the possibility that harmful materials may be present in on-site drinking water. The objective of the study reported here was to characterise drinking water at Hanford Site DOE facilities and to compare current levels of contaminants to state and federal standards for water quality. This information may be relevant to site workers and to planners involved with determining future use of the site.

Regulatory Oversight

Under the federal Safe Drinking Water Act (SDWA) and its 1986 and 1996 amendments, U.S. EPA set legal limits on the levels of certain contaminants in drinking water. U.S. EPA also established water-testing schedules and methods that water system owners must follow. The state governments, through their health departments and environmental agencies, were expected to accept the major responsibility for the administration and enforcement of the regulations set by U.S.

EPA. In the state of Washington, the federal drinking-water laws are implemented by the Washington State Department of Health (WDOH) through state administrative codes (Washington Administrative Codes [WAC]). All water systems on the Hanford Site are designated by Washington State as public water systems and are operated in accordance with the requirements contained in the applicable state codes. Water quality at Hanford is monitored according to written contaminant-specific plans that are in keeping with both state and federal drinking-water laws and are acceptable to WDOH.

Description of Hanford Site Drinking-Water Systems

The Hanford Site historically was divided into operational areas. The six 100 Areas (B/C, KE/KW, N, D, H, and F) on the shore of the Columbia River are where the nine retired plutonium production reactors are located. The 200-East and 200-West areas include facilities that processed the irradiated materials from the reactors. Fuel rods for the reactors were manufactured in the 300 Area, near the city of Richland. The 400 Area is the location of the Fast Flux Test Facility (FFTF), a liquid-sodium-cooled test reactor built by DOE in the 1970s to test equipment and fuel for the Liquid Metal Fast Breeder Reactor Program. The remainder of the site, the 600 Area, houses facilities that serve the entire site or more than one specific production area, and includes several non-DOE commercial and research facilities.

Four principal drinking-water pumping facilities were used on the site in 2001 and 2002. Most site workers were provided with drinking water pumped from the Columbia River. The 100-B Area pump house served as the primary

Columbia River pumping station, and the pump house in the 100-D Area functioned as its emergency backup. Together these two facilities potentially supplied water to an average of 4,184 people per workday in the 100, 200-East, 200-West, and 600 areas.

Water for the 100-K Area was obtained via its own river pump house and was potentially served to an average of 631 people per day. Drinking water in the 400 Area was pumped from three wells of varying depths located within the 400 Area complex.

Generally, only one of the three wells is operated at a time, but water from more than one well can be commingled in a holding tank prior to dissemination through the distribution system. An estimated 389 people were potentially served each workday from these wells. In the 300 Area, an average of 1,805 people was served each day, but the water for this area was supplied by the city of Richland.

All drinking water obtained from the Columbia River was filtered and chlorinated at treatment plants on the site. Active river water treatment plants were located in the 100-K, 100-N, and 200-West areas. Multiple treatment processes were provided in series to remove turbidity and microorganisms, and then to inactivate or remove a specific percentage of Giardia cysts or other pathogenic organisms.

At a minimum, filtration was provided at each river water treatment plant to achieve the required reduction in turbidity. Turbidity was monitored continuously to maintain performance expectations. Water from the 400 Area wells was chlorinated but not filtered. Drinking water in the 300 Area was delivered to consumers via a DOE-owned and -operated distribution system and was chlorinated on site when needed even though the water had already been treated by the city of Richland.

Sampling and Analysis Information

Hanford Site drinking-water systems have been classified by the state of Washington as Group A public water systems. Group A systems can be either community or noncommunity systems.

The systems at Hanford are defined as nontransient, noncommunity systems. State and federal laws requiring the monitoring of radiological contaminants in drinking water apply to community systems and therefore are not directly applicable to the Hanford Site. Radionuclides in DOE systems at Hanford are, however, monitored to community system requirements to comply with the requirements of DOE Order 5400.5, a DOE Directive. Monitoring includes collecting and analysing samples and comparing analytical results with established state and federal drinking-water standards and guidelines.

Radiological Sampling

Drinking-water samples for radiological analysis were collected and analysed quarterly in 2001 and 2002. Untreated river water samples were collected at the pump houses in the 100-B, 100-D, and 100-K areas in 2001. Well water samples collected during 2001 in the 400 Area consisted of treated water collected at a consumer-accessible sink. In 2002, in compliance with recently revised drinking-water regulations, all samples were collected after the water was treated but before it was distributed to the consumer. Collection sites in 2002 included treatment plants in the 100-N, 100-K, 200-West, and 400 areas. A state-certified analytical laboratory in Richland, Washington, analysed all samples during both years for gross alpha, gross beta, tritium, and strontium-90. Gross alpha and gross beta measurements are general indicators of radiological contamination. Tritium (12.3-year half-life) and strontium-90 (28.8-year half-life) are materials that were produced in large quantities at Hanford and are also components of atmospheric fallout from worldwide nuclear-weapons testing. Tritium also is produced naturally in the atmosphere.

To comply with revised regulations, one sample from each water supply in 2002 also was analysed for iodine-131, radium-226, and radium-228. Iodine-131 is a manmade isotope produced by the detonation of nuclear weapons and by nuclear-reactor operations. Because iodine-131 has a short half-life (eight days), it is not likely that it would be measurable in Hanford Site drinking water obtained from either river water or groundwater sources. Both radium-226 (1,599-year half-life) and radium-228 (5.8-year half-life) are isotopes that result from the decay of naturally occurring uranium-238 and thorium-232, respectively. Radium should be measurable in both surface and groundwater at Hanford, but U.S. surface waters such as the Columbia River usually have low radium levels.

A continuing program of internal laboratory quality control, participation in inter-laboratory crosschecks, replicate sampling and analysis, submission of blind standard samples and blanks, and splitting of samples with other laboratories verified the quality of radiological results. These verifications in 2001 and 2002 were summarised and published in the Hanford Site's annual environmental reports.

Radiological Results

Thirty-two drinking-water samples were analysed for gross alpha in 2001 and 2002 and alpha activity was detected only in one sample. This river water sample was collected in 2001 at the 100-D Area pump house. Gross beta was detected in one of 24 river water samples analysed during 2001 and 2002 and the measured concentration was about 8.4 percent of the U.S. EPA screening level of 1.85 becquerels per litre (Bq/L). This sample was collected at the 100-

N Area water treatment plant, which gets water from either the 100-B Area pump house or the 100-D Area pump house. All eight drinking-water samples from the 400 Area (groundwater) analysed during 2001 and 2002 were positive for gross beta, with the highest concentration measured during 2001.

Strontium-90 was measured in 23 of the 24 samples from river water sources analysed during 2001 and 2002. The highest concentration, about 0.7 percent of the U.S. EPA limit of 0.296 Bq/L, was seen in water from the 100-D Area pump house in 2001. The 100-D Area intake is located downstream of the 100-B, 100-K, and 100-N areas, where groundwater containing strontium-90 is found near the river.

Strontium-90 was not detected in any of the 400 Area well water samples in either year. Tritium was detected in five of 11 river water samples analysed in 2001 and in two of nine river water samples analysed in 2002. The highest concentration (~1.2 percent of the U.S. EPA annual average limit) was measured in water from the 100-B Area pump house. This drinking-water intake is the one farthest upstream at the site and the one least likely to be influenced by Hanford contaminated effluents entering the river.

The highest tritium concentration measured in drinking water on the site in 2001 and 2002 was found in a groundwater sample from the 400 Area. This concentration was 19 percent of the U.S. EPA annual average limit for tritium in drinking water and was slightly more than 16 times greater than the highest level measured in river water samples. Annual average tritium concentrations in 400 Area drinking water in 2001 and 2002 were 117 Bq/L and 128 Bq/L, respectively.

A groundwater tritium plume originating from the former plutonium-processing facilities in the 200 areas is known to affect well water in the 400 Area, and tritium levels in 400 Area drinking water can fluctuate depending on which of three drinking-water wells is being used. Tritium concentrations are usually lowest in the deepest well, which is considered the primary supply well.

Tritium concentrations above the annual average U.S. EPA limit of 740 Bq/L have been measured in the two shallowest wells in recent years, and the annual average level in 400 Area drinking water could potentially exceed the U.S. EPA standard if either of these two wells were used as the primary water source for most of the year. Neither of these wells was used in 2001 and 2002, however, and when water from one of these wells is needed, the current DOE policy is to use the well with the lowest tritium level as demonstrated by sampling and analysis.

Iodine-131, radium-226, and radium-228 were monitored once at all four sampling locations in 2002. As expected, iodine-131 was not detected in any samples. Radium-226 was, however, found in all four samples analysed during the year. The maximum radium-226 concentration was measured in Columbia River water from the 200-West Area treatment plant.

Radium-228 was found in two of three river water samples but not in the 400 Area well water sample. The highest detectable radium-228 level was found in water sampled at the 100-N Area treatment plant. As stated earlier, both the 100-N Area and 200-West Area treatment plants obtain water from the 100-B Area pump house or from the 100-D Area pump house if the 100-B Area facility is off line. There are no established drinking-water limits for radium-226 and radium-228 individually, but U.S. EPA has established a limit for the two combined. The highest concentration of radium-226 and radium-228 combined was measured in two river water samples and was only 2.2 percent of the MCL of 0.185 Bq/L.

Nonradiological Sampling

Drinking-water samples collected for non-radiological analyses were obtained in accordance with the WDOH-approved plans and procedures of the Hanford Site. Samples for coliform analyses were collected each month at representative locations around the site (primarily building sinks) by qualified personnel and transported to the analytical laboratory at the Benton-Franklin Health District office in Richland, Washington, for analysis. Disinfection requirements for all water sources included ensuring sufficient water and disinfectant contact times, maintaining a free-chlorine residual at each treatment plant prior to supplying consumers, and maintaining a detectable free-chlorine residual within each distribution system.

The chlorine residual was monitored continuously at the water treatment plants, and samples were collected daily from each distribution system. Monthly samples were collected in the 100-K, 100-N, 200-East, 200-West, 400, and 300 areas for coliform and total organic carbon analyses (surface-water systems only), and quarterly or annually (400 Area) at several locations within each distribution system to monitor levels of undesirable halogenated organic by-products (trihalomethanes [THMs] and haloacetic acids [HAA5s]).

Inorganic- and orga nic-chemical monitoring of Hanford water systems was performed in accordance with federal and state monitoring schedules. Samples for a suite of volatile-organic-compound (VOC) analyses were collected annually from the 100-N, 100-K, and 200-West areas, and once every three years from the 400 Area. Semivolatile organic compounds (SOCs) were monitored every year at the 200-West, 100-N, and 100-K areas and every three years in the 400 Area. Both the VOC and SOC samples were monitored for a wide variety of U.S. EPA-regulated and state-unregulated materials. Inorganic contaminants, including nitrate, fluoride, arsenic, and selected metals, were monitored annually in the 100-N, 100-K, 200-West, and 400 areas. Samples for lead and copper analyses were collected every three years at the above locations and in the 300 Area, but were not collected during 2001 and 2002. State-approved site water purveyor procedures were followed for the collection and timely transport of the samples. Chain-of-custody control provided further insurance that the requirements were met for documenting and maintaining custody of the samples from their point of origin to receipt at the laboratory.

Nonradiological Monitoring Results

In 2001 and 2002, all of the river water treatment plants effectively removed or inactivated 99.9 percent of Giardia lamblia and 99.99 percent of viruses. None of the Hanford Site drinking-water samples analysed for total coliform bacteria in 2001 and 2002 were found to be coliform positive. Concentrations of natural organic matter (disinfectant by-product precursors) in drinking-water samples were in compliance with regulatory requirements. Sampling also showed that annual average concentrations of undesirable halogenated organic by-products (HAA5s and THMs) were below their respective MCLs of 60 [micro]g/L and 80 [micro]g/L. No U.S. EPA-regulated VOCs were detected by the analytical lab in either year.

Scans for SOCs in surface water and groundwater systems were conducted in early 2002, and results were reported for 85 regulated and unregulated materials. Only one contaminant (Dalapon [2,2-dichloropropanoic acid]) was detected, and it was seen in water from the Columbia River. The highest amount detected (0.97 [micro]g/L) was far below the MCL of 200 [micro]g/L established by U.S. EPA. Dalapon is a herbicide that is used to control grasses in a wide variety of crops. It is also used in noncrop applications in lawns, in drainage ditches, along railroad tracks, and in industrial areas. This chemical was used on the site in the mid- to late-1990s, but it is believed that the concentration measured in 2002 is related to applications of the chemical to orchards and farmlands located upstream of the site.

Potential inorganic contaminants monitored at the site in 2001 and 2002 included 27 state- and U.S. EPA-regulated materials and parameters. Analytical results indicated that the concentrations of all regulated metals and other inorganic contaminants in site drinking-water samples collected in both years were below their respective maximum concentration limits. Sanitary surveys of Hanford's Group A water systems were performed by the Washington State Department of Health in June 2001. Each survey included an on-site review of the water source, facilities, equipment, operation, and maintenance of the water systems to evaluate their ability to reliably produce and distribute safe drinking water. All systems were found to be well operated and maintained, and monitoring program plans were satisfactory.

Concentrations of all radionuclides in drinking water supplied to Hanford Site facilities by DOE were below state and federal regulatory limits in 2001 and 2002. Tritium and gross beta levels in 400 Area drinking-water wells were elevated relative to other water sources on the site, and tritium concentrations in the two shallowest wells can potentially be high enough to exceed the 740 Bq/L annual average drinking-water standard. DOE's practice of using the two shallow wells only when needed and then using the well with the lowest tritium level as determined by sampling and analysis would, however, keep the annual average concentration below the regulatory limit for community systems.

Because of its upstream location, water pumped from the river at the 100-B Area pump house is not likely to contain contaminants from the fugitive groundwater discharges entering the river along the Hanford Reach shoreline. Water pumped from the river at the 100-K, 100-N, and 100-D areas could potentially contain contaminants from shoreline groundwater discharges that occur in or near each of the upstream reactor areas. Studies have shown, however, that the highest contaminant concentrations are localised near shoreline discharge points and that concentrations in river water samples collected outside the immediate discharge zones are below the limits set by state and federal drinking-water standards (e.g., Poston et al., 2003). Estimated groundwater discharge volumes throughout the Hanford Reach are small (~0.77 [m^3] per second) compared with the total flow of the river (annual average river flow in 2002 was 3,340 [m^3] per second), and shoreline contaminant concentrations are quickly diluted to very low levels.

Concentrations of the specific radionuclides monitored in Hanford Site drinking water obtained from the Columbia River in 2001 and 2002 were similar to concentrations measured in river water samples collected downstream of the site at the city of Richland's drinking water pump house. In most cases, measured contaminant levels in river water samples were very low, and in all cases they were below state and federal annual average drinking-water limits.

The authors' monitoring results indicated that current filtering systems and disinfectant applications are effectively controlling populations of potentially harmful microorganisms and that most VOC, SOC, and inorganic contaminants do not affect drinking water supplied to Hanford Site facilities. Proper operation and maintenance of the distribution systems is one factor that protects Hanford Site drinking-water consumers from microbiological contaminants. Personnel operating the water treatment plants and distribution systems are required to be certified by WDOH as being knowledgeable about water treatment processes and technologies and to have proven themselves, by examination, capable of effectively and safely operating the water system. All public water systems on the Hanford Site, both production and distribution, are operated by WDOH-certified Water Works Operators who maintain certifications at levels commensurate with the complexity of the systems.

Most nonradiological contaminants detected in Hanford Site drinking water were by-products of the disinfection (chlorination) process. An unrelated chemical contaminant noted in 2001 and 2002 was Dalapon, an herbicide used for agricultural weed control off the site.

Concentrations of this chemical were well below the MCL established by U.S. EPA for this material. With the closing of the production reactors and associated plutonium separation facilities at Hanford, the federal government began focusing on waste cleanup, waste management, and environmental remediation and restoration.

Current DOE goals include reducing the size of the Hanford Site from 1,517 [km^2] to 194 [km^2] by 2012 and preparing portions of the cleaned-up site for other DOE missions, non-DOE federal missions, and other public and private sector uses (DOE, 2002). A recent Presidential Proclamation (Establishment of the Hanford Reach National Monument, 2000) placed ~800 [km^2] of the site in the Hanford Reach National Monument under management by DOE and the U.S. Fish and Wildlife Service. The results indicate that safe, high-quality drinking water will be available at the tap as long as Hanford-specific procedures and standard industry practices acceptable to WDOH continue to be followed.

Challenges for Controlling Water Discharges and Aquatic Invasive Species

In January 1991, the Ministry of Health in Lima, Peru, started receiving reports of an illness later identified as cholera. Over the course of the next several years, the epidemic spread to all but one Latin American country, infecting at least 100,000 and killing approximately 10,000 individuals. Evidence indicates ballast water from ships moving between Asia and South America was the most likely source of the outbreak—the first in the Western hemisphere in over a century. Ultimately, the cholera outbreak cost Peru $770 million, primarily as a result of trade embargoes on food and decreased tourism. Unfortunately, the cholera epidemic in Peru is just one example of a non-native species causing harm internationally to the environment, human health, and the economy. Non-native species are species of plants, animals, and microbes that do not occur naturally in a particular habitat.

Approximately 50,000 non-native species are located in the United States. Some non-native species, including corn, wheat, rice, poultry, and cattle, are utilised for agriculture and other purposes in the United States. However, when the species are introduced into new environments, their natural predators are often absent and the species "can compete with native biota; displace them; predate upon them; parasitise and transmit or cause diseases; reduce growth and survival rates; cause decline, extirpation (local extinction) of populations, or extinction." These invasive species are a destructive subset of non-native species that cause environmental damage and lead to adverse economic consequences in agriculture, forestry, and other industries. Internationally, governments and other entities have already spent billions of dollars to remove and control invasive species. Invasive species also cause damage to the environment and other species, with "approximately 35-46 percent of the species on the endangered species list [present either] partly or entirely because of the effects of invasive species."

The International Maritime Organisation (IMO) considers invasive marine species to be one of the four greatest threats to the world's oceans. However, "unlike other forms of marine pollution, such as oil spills, where ameliorative action can be taken and from which the environment will eventually recover, the impacts of invasive marine species are most often irreversible."

In marine environments, ballast water from ships is the primary mechanism for distribution of species between aquatic ecosystems. Ships often introduce invasive species into new habitats when releasing ballast water, which vessels take on for balance and stability. Despite the tremendous risk posed internationally to the economy, public health, and the environment, current regulations for ballast water are limited in scope and effectiveness.

Recently, in Northwest Environmental Advocates v. US. Environmental Protection Agency, the Ninth Circuit invalidated a. long-standing Environmental Protection Agency (EPA) regulation that exempted all vessel discharges from permitting requirements under the Clean Water Act (CWA).

In light of the potential repercussions from Northwest Environmental Advocates, this Chapter argues that the reactions by EPA, Congress, and the states to the Ninth Circuit's decision have the potential to trigger the adoption of ballast water treatment technologies on an international scale. Further, this Chapter proposes that the responses to Northwest Environmental Advocates could bolster efforts to ratify the International Convention for the Control and Management of Ships' Ballast Water and Sediments, a binding international framework adopted in 2004 by the IMO, by driving the adoption of treatment technologies.

It begins with an explanation of the problems created by untreated ballast water and provides an overview of the current status of treatment technologies. It describes the current international regimes that address invasive species and ballast water discharges, and argues that current, binding standards are insufficient to drive adoption of the technology necessary to prevent introductions of aquatic invasive species.

It further analyses the lack of effective controls for ballast water in the United States and asserts that EPA's ineffective response to the decision in Northwest Environmental Advocates could spur states to adopt individual and regional frameworks that require ships to install effective ballast water treatment systems. The Chapter claims that because individual states in the United States are taking the initiative to develop technology requirements that are as stringent, if not more stringent, than the IMO convention, the United States could still trigger the adoption of effective treatment technologies on an international scale. Finally, the Chapter offers suggestions on how the United States could reconcile the conflicts between state, federal, and international regulatory mechanisms.

Ballast Water as a Conduit for Invasive Species Transfers

Ships use ballast to ensure safe operation at sea. Ballast is "any solid or liquid placed in a ship to increase the draft, to change the trim, to regulate the stability, or to maintain stress loads within acceptable limits." The uptake and discharge of ballast water leads to the inadvertent transport of thousands of organisms around the globe on a daily basis. This Part begins with an overview of the harm caused by aquatic invasive species and ballast water discharges and ends with a review of the current status of ballast water treatment technologies.

Dispersal of Invasive Species Through Ballast Water Discharges

Before the late 1800s, ships typically carried heavy materials such as rocks and sand as ballast. Today, ships pump water into holds to maintain balance, to relieve stress on the ship, and to fulfill other functions required for safe operation of the vessel. Ships typically load ballast water when discharging their cargo, carry the ballast water while their cargo holds are empty or low, and discharge the water when filling the cargo hold. The loading and unloading of the ballast water in relation to ship cargo allows the ship to maintain equilibrium in the vessel's stability.

The shipping industry oversees the transport of over 90% of the world's goods and commodities and is the primary distributor of aquatic invasive species around the world. Overall, experts estimate that shipping is responsible for approximately 80% of the invertebrate and algae introductions into North America, with ballast water and hull fouling accounting for 90% of the introductions. As such, ballast water "is widely regarded as the leading modern-day vector of marine bioinvasions." Shipping vessels transport over ten billion tons of ballast water each year, with each vessel containing anywhere from "several hundred litres to more than 100,000 tons, depending on the size and purpose of the vessel." While the use of water for ballast is efficient and convenient for ships, the water typically carries a wide variety of small species, "including bacteria, microbes, small invertebrates, and the eggs and larvae of larger species." Ballast water carries an estimated 7000 different species of animals and plants around the world each day. Although the majority of these hitchhiking species do not survive the harsh conditions during their ballast water journey, those species that do survive can "become invasive, out-competing native species and multiplying into pest proportions."

Ballast water discharges are the source of many of the most damaging invasive species problems around the world. One famous example of the threat posed by invasive species in ballast water is the damage caused by the zebra mussel in the Great Lakes. Although scientists do not know the exact time and place of the first zebra mussel establishment, evidence indicates that ballast water discharges caused the first zebra mussel introduction into Lake St. Clair and the Great Lakes in the 1980s. By 1993, zebra mussel populations stretched from Quebec to Louisiana. Recently, states discovered zebra mussels as far west as Lake Mead in Nevada. Additionally, the first confirmed zebra mussel in California appeared in early 2008. Within twenty years, scientists expect the zebra mussel to infest most freshwater habitats throughout the nation, thereby reducing native mussel populations by over fifty percent and causing the extinction of up to 140 species. The "cost of Zebra mussel prevention and remediation just in the Great Lakes is estimated to be between $100 million and $400 million per year."

The zebra mussel is only one example of the many harmful species that ships have transferred into new ecosystems through ballast water. Other examples of non-native species invasions caused by ballast water abound. For example, in the Black Sea, Leidy's comb jellies—a native to the East Coast of North and South America—invaded the water in densities as high as 500 jellies per cubic metre, and scientists expect the species to spread to areas outside of the Black Sea. With a voracious appetite, a lack of predators, and a reproductive system that could produce 8000 eggs in a day, the comb jelly "takeover" caused the collapse, and sometimes elimination, of fish stocks in the Black Sea; the comb jelly assault "is thought to have cost Black Sea fisheries $30 million a year directly and is the critical factor in a crisis that has deprived some 2 million people of their livelihood, either in the fishing fleets or in the businesses that served them."

Attempts to eradicate aquatic invasive species are both expensive and relatively ineffective. Typically, once an aquatic invasive species establishes a population in a new environment, the species becomes nearly impossible to remove and there are few international examples that demonstrate successful eradication Of aquatic invasive species. The costs associated with eradicating an aquatic invasive species are tremendous, even when the species has yet to spread to unconfined areas. For example, over the course of five years,

California spent roughly $7 million to remove the Mediterranean green seaweed from two areas in Southern California, despite the fact that "the infestations were detected while they were still confined." Because of the costs and difficulty in removing introduced aquatic species, a proactive—rather than reactive—stance is vital to preventing species introductions.

However, since ballasting is necessary for safe operation of ships, "the only effective way to stop the spread of unwanted organisms is to prevent them [from] being dumped in foreign ports." As global trade increases, invasions by non-native species are also likely to increase, since the rate of invasion in coastal ecosystems correlates with the volume of ship traffic. To date, however, industry has not adopted treatment technologies on a widespread basis. This Part continues with an overview of the current state of ballast water controls and technologies.

The Technology Gap

Despite the problems with ballast water and aquatic species introductions, a gap remains between the development and the adoption of economically viable, effective treatment technologies. The majority of commercial vessels currently use ballast water exchange as a means of minimising the distribution of viable aquatic species. Ballast water exchange involves the replacement of ballast water taken on during port or coastal visits with open ocean water, or vice versa. Ballast water exchange relies on the biological principle that the ocean environment is generally inhospitable for species adapted to living in coastal regions, and oceanic organisms are unlikely to survive if discharged into freshwater, estuarine, or inshore coastal waters.

Ballast water exchange provides an incomplete solution, however, to the problems with ballast water. Studies of ballast water exchange indicate that its effectiveness is "highly variable," and ballast water exchange "as currently practiced probably has little effect in reducing the introduction of planktonic non-indigenous species." For example, in one study, even after three tank exchanges, only ninety-five percent of the original ballast water had been replaced. As a result, ballast water exchange "can rarely, if ever, remove all original organisms from ballast water." Additionally, ballast water exchange may pose a threat to the ship and the safety of the crew by creating instability and other dangerous forces on the hull of the ship. Due to the lack of effectiveness and the overall safety threats, national and international governments should view ballast water exchange as an interim measure that is unsuitable in the long term for eliminating ballast water introductions.

In light of the problems with ballast water exchange, companies are exploring a wide range of treatment options for ballast water discharges. Yet, "until recently, financial investment in the research and development of ballast water treatment systems has been limited and the advancement of ballast water treatment technologies slow." Companies have explored the possibility of implementing both shoreside and ship-based treatment facilities. Although experts have examined options for shore-based treatment of ballast water, which involves transfer of water from a ship to a treatment facility or vessel, the option presents several challenges.

Among other problems, new treatment facilities are necessary for shoreside treatment because "current shoreside wastewater treatment plants are not equipped to treat saline water," ports and ships would need to retrofit their facilities to allow shoreside discharges, and, while technically feasible, shoreside facilities or vessels would be costly. Moreover, "the sheer number of ships entering large port cities daily prevents major ports from instituting effective shore-based solutions."

Thus, the focus over the last several years has been on shipboard treatment systems, which are likely to result in greater flexibility for vessels and fewer delays than shoreside treatment facilities. Shipboard treatment options generally fall into one of five categories: 1) mechanical operations, such as filtration or cyclonic separation; 2) chemical treatment with substances such as biocides, chlorine, hydrogen peroxide, and sodium; 3) physical treatment through heat, ultraviolet, ultrasonic, or deoxygenation technologies; 4) biological treatment, which involves the introduction of organisms, such as yeast, that kill or create conditions that kill other species; or 5) a combination of these technologies. In a recent evaluation by California of the current state of shipboard treatment systems, the results for these technologies appeared promising. For a wide range of tested organism sizes, the results indicated that over half of the technologies met California's performance standards for ballast water discharges—the most stringent in the world. Additionally, two of these technologies, which complied with every California performance standard, are already considered "commercially available."

Despite the current potential of shipboard technologies to reduce or eliminate introductions of aquatic invasive species, the shipping industry has yet to move away from ballast water exchange and toward treatment technologies. Historically, several barriers hindered the development and adoption of new technologies for commercial vessels. Some of the major barriers included "equipment design limitations, the costs of technology development, and the lack of guidelines for testing and evaluating performance." However, the primary deterrent for investors, the shipping industry, and technology developers to invest in new technologies has been "the absence of a specific set of ballast water performance standards." Without a specific set of standards, both developers and industry representatives risk spending large sums of money to implement technologies that may later prove obsolete due to new or conflicting federal, state, or international requirements.

The resulting dynamic between the availability of technology and the creation of binding performance standards appears to create a circular "chicken and the egg" dilemma—industry refuses to adopt existing technology because of the lack of regulatory certainty, while international and domestic governments avoid adopting rigid timelines for standards because of concerns about the technologies' availability. The dilemma begs the question of what, if anything, can break the current dynamic between industry and various governments.

The International Framework for Addressing Ballast Water Discharges

The distribution of invasive species through ballast water has local, national, and international ramifications. Because the prevention of species transfers is the only viable solution for halting future aquatic invasions, this Chapter asserts that a federal and international regulatory framework is necessary to control and minimise introductions of aquatic invasive species. Despite the economic and ecological risk posed by ballast water discharges, however, there are few regulations internationally or in the-United States that directly and comprehensively address the threat. Thus, a gap remains between the development and implementation of ballast water technologies. Part III begins with a discussion of the current international framework for addressing ballast water discharges and analyses whether the International Convention for the Control and Management of Ships' Ballast Water and Sediments—the Ballast Water Convention—could trigger the adoption of effective ballast water treatment technologies.

Deficiencies in Current International Law for Addressing Ballast Discharges

A basic understanding of the deficiencies in current international law is necessary to understand the need for the adoption of the Ballast Water Convention, the first international framework for addressing ballast water discharges through binding performance standards. No binding international treaty currently exists that directly addresses the problem of invasive species in a comprehensive manner, and few countries has created domestic policies. Although at least a dozen treaties mention invasive species introductions, the treaties do not provide the specific enforcement or regulatory mechanisms necessary to implement effective invasive species controls. Additionally, the majority of treaties only "focus on a specific dimension of alien-related issues, with regard to a particular protection objective (e.g. migratory species), kind of activity (e.g. introductions for aquaculture) or potentially damaging organism (e.g. pest)." Because many of these treaties and conventions are binding upon their signatories, "it is [the parties'] duty to adhere to [treaty] provisions to the best of their ability." Nevertheless, a "fundamental conundrum" is created by the gap between the party's duty to adhere to the agreement, and the presence of an effective mechanism for implementing specific obligations. The following discussion reviews the gaps in international regulation left by the two most important binding legal instruments currently in force for addressing invasive species—the Convention on Biological Diversity and the United Nations Convention on the Law of the Sea—and illustrates the need for a binding international agreement to deal specifically with ballast water discharges.

The Convention on Biological Diversity

One of the most important binding international instruments to address invasive species problems is the Convention on Biological Diversity (CBD). The CBD is "the only globally applicable, legally binding instrument to address generally alien species introduction, control and eradication across all biological taxa and ecosystems." There are currently 191 parties to the treaty, which does not include the United States. Of key importance is Article 8(h) of the CBD, which requires that "as far as possible and appropriate," each contracting state should "prevent the introduction of, control or eradicate those alien species which threaten ecosystems, habitats or species." The Convention lists Article 8 as part of a list of "in-situ conservation" measures, which are defined in the CBD as "the conservation of ecosystems and natural habitats and the maintenance and recovery of viable populations of species in their natural surroundings and, in the case of domesticated or cultivated species, in the surroundings where they have developed their distinctive properties."

Although the CBD is binding upon signatories and indicates that parties should both prevent and eradicate threatening alien species, the Convention's broad language gives little guidance to parties on how to implement or enforce the provisions of the Article. In addition, the broad "as possible and appropriate" language of Article 891 creates a loophole that might allow states to avoid implementation of effective invasive species policies; based on the text, a state hoping to avoid any obligation to prevent or eradicate an alien species could deem the action inappropriate or impossible, and thereby avoid implementing key measures. As a result, Article 8 alone provides a weak foundation for building a consistent and comprehensive international policy on aquatic invasive species. In recognition of the "gaps and inconsistencies in the international framework" for dealing with invasive species, the contracting parties to the CBD asked the Subsidiary Body on Scientific, Technical, and Technological Advice (SBSTTA) to develop guiding principles for the effective implementation of Article 8. In 2000, the SBSTTA issued interim guiding principles, and the Conference of the Parties (COP) requested that the international community implement the principles. In March 2001, the SBSTTA finalised the guidelines—Guiding Principles for the Implementation of 8(h) —and in 2002, the COP endorsed the guidelines.

The COP intended to provide "guidance for developing effective strategies to minimise the spread and impact of invasive alien species," and to "give Governments clear direction and a set of goals to aim toward." The SBSTTA guidelines consist of fifteen guiding principles, which address application of the precautionary approach, research and public education efforts, prevention, and mitigation. The guidelines also outline the factors countries should consider when addressing invasive species problems and provide examples of how individuals can introduce invasive species.

Although the SBSTTA guidelines touch on various aspects of a comprehensive invasive species policy, the guidelines do not provide substantive standards, nor are they binding on CBD parties As a result, the guidelines provide only a broad overview of the elements a state could address in an invasive species framework, Because of the lack of detailed implementation and enforcement mechanisms for ballast water discharges, the guidelines are likely to lead to inconsistencies in how various countries implement controls on ballast water discharges. Without additional, detailed policy requirements under the CBD, or a binding framework to guide substantive aspects of an aquatic invasive species policy, the CBD provides a weak foundation for driving the adoption of treatment technologies and effective performance standards.

The United Nations Convention on the Law of the Sea

More specific to introductions in marine and coastal ecosystems is the United Nations Convention on the Law of the Sea (UNCLOS)) UNCLOS "provides the most comprehensive and significant document to date dealing with the prevention of marine pollution, among other matters concerning the world's oceans." Enough countries have signed and ratified the treaty that even countries that have not ratified the treaty, including the United States, comply with UNCLOS. The treaty specifically addresses invasive species in Article 196, which indicates:

States shall take all measures necessary to prevent, reduce and control pollution of the marine environment resulting from the use of technologies under their jurisdiction or control, or the intentional or accidental introduction of species, alien or new, to a particular part of the marine environment, which may cause significant and harmful changes thereto.

Although the Article invokes the requirement for states to "take all measures necessary" to prevent species introductions, the Article lacks details on how states are to achieve or enforce this objective. Standing alone, Article 196 appears to classify invasive species as a type of pollution. By including a reference that ties together pollution and invasive species—namely, that states are to "control pollution of the marine environment resulting from ... the intentional or accidental introduction of alien, species or new"—Article 196 on its face indicates that invasive species could be a type of pollution under UNCLOS. Despite this implied inclusion of invasive species in the definition of pollution in Article 196, there is no indication that invasive species fall under other articles of UNCLOS that expressly address pollution from vessels and provide enforcement authority to states. For example, Article 211, Pollution from Vessels, outlines provisions for addressing operational pollution caused by ships. Article 211 requires states to "adopt laws and regulations for the prevention, reduction and control of pollution of the marine environment" from vessels and authorises states to adopt laws within their exclusive economic zone (EEZ) for enforcement purposes, Ideally, Article 196 would link ballast water and invasive species discharges to the provisions in Article 211 for vessel pollution, or would provide similar details on enforcement and legal authority over vessel discharges. Unfortunately, although Article 211 plays an important role in allowing states to implement and enforce vessel pollution controls within their EEZ, there does not appear to be a clear link between Article 196's discussion of alien invasive species and Article 211's discussion of vessel pollution. Had Article 196's invasive species provision been clearly linked to Article 211's vessel pollution provisions, UNCLOS might have provided a basis for the adoption of ballast water controls and enforcement mechanisms to address invasive species introductions.

Because Article 196 is not linked to Article 211's vessel pollution provisions, and because UNCLES contains few, if any, details on states' obligations and powers for preventing the introduction of alien species, UNCLES is unlikely to serve as a strong catalyst for the adoption of stringent or uniform international controls for ballast water. The lack of enforceable details in Article 196 is also likely to provide a loophole for countries that want to avoid their obligations under UNCLES. Further, the lack of specific mandates under UNCLES for ballast water discharges is likely to lead to inconsistencies in how countries adopt regulatory mechanisms. Ultimately, without greater regulatory certainty for the shipping industry, and without additional mechanisms that force countries to adopt effective treatment standards, it is doubtful that UNCLES could serve as a catalyst for adoption of treatment technologies on an international scale.

The Ballast Water Convention as a Binding Regulatory Mechanism

Over the past decade, the IMO developed mechanisms to address the introduction of invasive species through ballast water. In 1992, the United Nations Conference on Environment and Development called on the IMO to address the transfer of organisms by ships. Although the IMO initially adopted guidelines for minimising the transfer of harmful aquatic organisms in 1993, it was not until 1997 that the IMO Assembly adopted by resolution the Guidelines for the Control and Management of Ships' Ballast Water, to Minimise the Transfer of Harmful Aquatic Organisms and Pathogens (IMO Guidelines).

The IMO Guidelines replaced earlier, less comprehensive guidelines and included provisions to assist interested parties "in minimising the risk of introducing harmful aquatic organisms and pathogens from ships' ballast water and associated sediments while protecting ships' safety." Measures recommended by the IMO Guidelines include minimising the uptake of organisms during ballasting by avoiding areas known to contain harmful organisms, cleaning ballast tanks, and avoiding unnecessary discharges of ballast water. Additionally, the Guidelines suggest ballast water management procedures; the procedures involve either exchanging ballast water as far away from shore as possible while complying with regional requirements, avoiding release of ballast water, or discharging the water into onshore facilities.

Because the IMO Guidelines were only voluntary, compliance was low for the specific, recommended practicess. As a result, the "voluntary guidelines resulted in differing rules among jurisdictions, complicating the shipping industry's efforts at compliance." Additionally, there were limitations on the approaches listed under the IMO Guidelines. For example, although ballast water exchange is currently one of the primary measures to minimise the risk of invasive species, ship safety often precludes the use of ballast exchange practices at sea. Further, because even repeated ballast water exchange does not remove all organisms from ballast water tanks, ballast water exchange is limited and variable in its effectiveness. Since ballast water exchange is only a step above a complete lack of management techniques, the IMO Guidelines are an incomplete and nonbinding attempt by the IMO to reduce aquatic species introductions. Recognising the limitations of the IMO Guidelines, the continuing problems with aquatic invasive species, and the need for binding targets for ballast water regulation, the IMO member countries agreed to develop a binding international framework to control ballast water discharges. In 2003, the IMO's Marine Environmental Protection Committee developed and finalised the draft text for the International Convention for the Control and Management of Ship's Ballast Water and Sediments (Ballast Water Convention. The IMO subsequently adopted the Ballast Water Convention by consensus at the IMO International Conference on Ballast Water Management for Ships in February of 2004. The Ballast Water Convention will enter into force twelve months after ratification by thirty states, representing 35% of the world merchant shipping tonnage. As of February 2009, there were eighteen ratifying parties, representing 15.36% of the world's shipping tonnage. Australia, Brazil, Finland, and the Netherlands are among the countries that signed, but have yet to ratify, the Convention as of March 2009.

Key Provisions of the Ballast Water Convention

The Ballast Water Convention is the first treaty seeking to "prevent, minimise and ultimately eliminate the risks to the environment, human health, property and resources arising from the transfer of harmful Aquatic Organisms and Pathogens through the control and management of ships'

Ballast Water and Sediments." The Ballast Water Convention contains twenty-two articles, regulations for implementation of the Convention's goals, and two appendices with sample reporting and certification forms. The Convention applies to flag ships of contracting parties, as well as ships "which operate under the authority' of a Party." This provision allows for "passive conformity as an increasing number of countries are forced to have their ships abide by the regulations for fear of financial losses." Additionally, "saving the Convention apply to those states under whose jurisdiction the ship operates can also be useful, in that this provision would require countries that may not even be signatories to the document to meet its requirements." Ultimately, universal compliance with the Ballast Water Convention could be achieved through port state jurisdiction, simply because the Convention's parties have the power to impose the Ballast Water Convention's discharge requirements within their jurisdictional waters, regardless of whether the flag country ratified the Convention.

Relying on Article 196(1) of UNCLES, the precautionary principle, and sustainable use principles, the Ballast Water Convention delineates a mandatory regime for ballast water management with the ultimate goal of eliminating ballast water's risks to the environment, avoiding "unwanted side-effects from that control and [encouraging] developments in related knowledge and technology." To implement the parties' ultimate goal, the Ballast Water Convention provides specific requirements for discharges of ballast water, including dates of compliance and ballast water exchange procedures.

For the ballast water exchange requirements, known as Regulation D-1 requirements, the Ballast Water Convention requires ships to conduct a ballast water exchange with at least a ninety-five percent rate of effectiveness. Whenever possible, this exchange must be conducted at least two hundred miles from shore in water at least two hundred metres deep. Exceptions to this discharge requirement include situations where the discharge is a threat to the safety or stability of the ship, or the operation would delay or deviate the ship.

Most importantly, the Ballast Water Convention sets binding performance standards, known as Regulation D-2 requirements, which limit the number of viable organisms allowed in ballast water discharges and ultimately replace the D-1 exchange requirements starting in 2011. Regulation D-2 limits the number of viable organisms per cubic metre that ships can discharge:

Ships conducting Ballast Water Management in accordance with this regulation shall discharge less than 10 viable organisms per. cubic metre greater than or equal to 50 micrometres in minimum dimension and less than 10 viable organisms per millilitre less than 50 micrometres in minimum dimension and greater than or equal to 10 micrometres in minimum dimension; and discharge of the indicator microbes shall not exceed specified concentration.

Additionally, Regulation D-2 limits the concentrations of "human health related" microbes, or indicator microbes, present in the discharge. The indicator microbes include cholera (Vibrio cholerae), E. coli (Escherichia coli), and intestinal enterococci. Regulation D-3 further indicates that all treatment technologies utilised by vessels are subject to approval by the IMO, and must meet IMO standards. The Convention distinguishes between technologies that employ or do not employ an "active substance," which the Convention defines as "a substance or organism, including a virus or a fungus, that has a general or specific action on or against Harmful Aquatic Organisms and Pathogens." To obtain approval, systems that do not use active substances must go through land-based testing and shipboard trials; once the trials are complete, the flag state issues a "type approval certificate," which indicates the technology meets the approval requirements of Regulation D-3 and can be utilised by flag states to meet the D-2 performance standards. Systems using active substances must clear additional hurdles to receive type approval. First, the system must receive basic approval from an advisory committee prior to beginning land-based testing and shipboard trials; this additional step is to ensure "that the use of the [active substance] poses no harm to the environment" and prevents companies from needlessly wasting research dollars on systems that damage the environment. Additionally, active substance technologies must receive final approval from the advisory body before a flag state can issue a type approval certificate. Typically, the approval process takes up to two years. As of 2008, only three systems hold type approval certificates, two of which utilise active substances.

The Convention provides a timetable for implementation of Regulation D-2 that depends on ship size and construction dates, and ultimately requires all ships to meet the D-2 performance standards by 2016. The performance standards do not provide a grandfathering clause for older vessels, which thereby "forces the entire fleet to shift technologies or practices" by the prescribed dates. Originally, the Convention required ships constructed in or after 2009 with a ballast water capacity of less than five thousand cubic metres to comply with D-2 requirements. However, because the Convention had yet to enter into force, the IMO General Assembly adopted a resolution that effectively delayed the requirements to no later than 2011 for ships built in 2009. To ensure compliance with the Ballast Water Convention's requirements, the Convention creates a binding obligation on vessels to keep detailed records of the ship's ballast water operations, and for each ship to develop a Ballast Water Management plan detailing how the ship will implement the Convention's provisions. The Convention also authorises parties to inspect the ship's ballast water certificate and record book, and to sample the ship's ballast watery When a state finds a ship is in violation of the Convention's requirements, the Ballast Water Convention authorises the state to take multiple actions.

First, the state under whose authority the ship is operating must establish sanctions for violations. If a state finds a ship in its waters to be in violation of the Convention, the state may then bring proceedings in its own court, or may furnish information and evidence to the flag state to show how the ship violated the Convention. Any sanctions imposed "shall be adequate in severity to discourage violations of this Convention wherever they occur." In addition to sanctions, the flag or port state may "warn, detain, or exclude the ship," and may prohibit the ship from discharging ballast water until the removal of any threats. By authorising states to test ballast water and bring enforcement actions for violations of the Convention, the Ballast Water Convention has the potential to maintain compliance among vessels.

In addition to vessel controls, the Convention also encourages parties to promote and facilitate research on ballast water controls. The Ballast Water Convention asks parties to make relevant information available for other parties who request information; parties can request information on other parties' scientific and technology programs, or on the development of ballast water technologies. To further the effectiveness of the program, the IMO, United Nations Development Programme (UNDP), member governments, and the shipping industry created a joint initiative known as the GloBallast Program. The primary purpose of the program is to assist less-industrialised countries with the adoption of ballast water controls and technologies. With a $10.2 million budget, the program has the potential to spur the adoption of a regulatory framework for ballast discharges in countries that might otherwise choose not to implement the provisions of the Convention. Overall, the "program is yet another example of the international cooperation and the determination of the international community to work together to achieve the Convention's objectives."

Finally, of key importance is a provision that allows states to adopt more stringent measures than those outlined in the Convention. As an active participant in the Convention's negotiations, the United States suggested that the Ballast Water Convention include a provision to preserve the sovereign right of a party to enforce more stringent measures than those adopted in the Convention. As written, Article 2 of the Ballast Water Convention preserves countries' sovereign right to adopt more stringent measures than those in the Convention:

Nothing in this Convention shall be interpreted as preventing a Party from taking, individually or jointly with other Parties, more stringent measures with respect to the prevention, reduction or elimination of the transfer of Harmful Aquatic Organisms and Pathogens through the control and management of ships' Ballast Water and Sediments, consistent with international law.

Although the United States might want more stringent performance standards in the Ballast Water Convention, the ultimate inclusion of this provision in the Convention should have the power to alleviate some of the concerns of parties who want to impose more stringent measures to protect their waters.

Moving Forward with the Ballast Water Convention

Overall, the Ballast Water Convention has the power to drive technology adoption by setting clear, unambiguous goals for adoption of ballast water treatment technologies. The Convention would also be a marked improvement from current international regulatory regimes, which are not comprehensive or clear with regard to ballast water and aquatic invasive species prevention, technology implementation, or enforcement. Ultimately, as a binding legal instrument, the Ballast Water Convention could provide an international floor for regulation and establish technology targets for the shipping industry. Further, since there is no grandfathering clause in the Ballast Water Convention, the Convention would force entire fleets of ships to adopt necessary treatment technologies.

The Convention, however, also has some significant problems. First, given the billions of tons of ballast water that ships discharge on an annual basis, the D-2 treatment requirements are unlikely to eliminate invasive species introductions. As purely a mathematical estimate, with a discharge rate that allows for up to ten viable organisms per cubic metre, and with individual ships containing as much as—if not more than—five thousand cubic metres of ballast water, vessels still could discharge tens of thousands of viable organisms during ballast water operations. Although it is difficult to predict which organisms will successfully establish invasive species populations, a high risk remains that species introductions will continue if countries adopt the Ballast Water Convention's performance requirements over more stringent requirements.

Some might also argue that the Convention's deadlines for adopting treatment technologies could be more stringent. Specifically, the Convention's deadlines treatment technology implementation could phase in technology requirements in fewer years than under the current deadlines. For example, ships built before 2009 with ballast water capacities of more than five thousand cubic metres do not need to meet the D-2 performance standards until 2016, but this performance standard could be tightened to 2012 or another year that is closer in time than 2016. Additionally, the Convention allows for the reexamination of treatment technologies and the timeframes located in Regulation B-3; just as the IMO previously extended the deadline for new vessels, the IMO could also extend the implementation dates for other ships. Through the Convention's review process, an IMO committee must periodically undertake a review to determine whether "appropriate technologies" are available for implementation of the D-2 standard and to examine all other aspects of the regulations. During the review of available technologies, the IMO must take into account safety issues, environmental impacts, compatibility of the technologies with ship designs, cost and economics, and the effectiveness of the treatment technologies in removing organisms and pathogens. Following the review, the committee can make recommendations for proposed amendments to the Convention's regulations. Through the review process, and under the broad language of the Convention, the IMO has the power to loosen the deadlines and other aspects of the Convention that drive technology adoption. As a result, although the mechanism could arguably be used to strengthen the performance standards, there is also the possibility that the IMO might weaken the performance standards by undermining the performance standards or the adoption timeline. By potentially allowing the IMO to delay or loosen treatment requirements, the review mechanism could also compromise the credibility of the Convention as a tool for forcing the adoption of treatment technologies.

Another key question is whether the Ballast Water Convention's enforcement provisions are sufficient to overcome the historical problems with enforcement under other international treaties. As exemplified by the International Convention for the Prevention of Pollution from Ships (MARPOL) and UNCLOS, international law tends to favour flag state enforcement over port state powers, resulting in imbalances in the implementation of enforcement provisions. Under MARPOL, port states and other parties are to provide the flag state with evidence that the ship violated the provisions of the Convention. The flag state is then obligated to investigate the violation and, if there is sufficient evidence, commence proceedings under the flag state's laws. Similarly, UNCLOS indicates that, except in exceptional circumstances or other situations specified under UNCLOS, ships are subject to the "exclusive jurisdiction" of their flag state on the high seas and the flag state has the power to preempt criminal proceedings of port states. Although "flag state jurisdiction is not essentially wrong," problems have arisen internationally because "flag states appear reluctant to enforce standards against their ships." In one study, out of three hundred violations reported by port states, flag states took action in only seventeen percent of the cases, potentially because the multinational corporations that own the vessels are often "more powerful than many flag states" and because flag states do not suffer from the direct environmental consequences of the vessels' actions.

Conversely, breaking with the frameworks adopted under MARPOL and UNCLOS, the Ballast Water Convention sets up a system that allows for both port and flag state enforcement. Unlike MARPOL's provisions for flag state enforcement and port state referrals, Article 8 of the Ballast Water Convention indicates port states can either initiate proceedings against the violating party using the port state's laws or may furnish information to the flag state; by allowing port states to initiate their own proceedings and pursue their own penalties, the Ballast Water Convention departs significantly from MARPOL and UNCLOS and has the power to alter some of the historical problems with enforcement. Additionally, port states have the power under the Ballast Water Convention, in addition to imposing sanctions, to "take steps to warn, detain, or exclude the ship." The Ballast Water Convention also goes beyond MARPOL by providing authority not only for inspections, but also for sampling of the ship's ballast water to determine if the vessel is in compliance with the performance standards or other regulatory requirements. Finally, because the Ballast Water Convention allows parties to "develop national policies, strategies or programmes" for ballast water management "with due regard to [the state's] particular conditions and capabilities," the Ballast Water Convention empowers port states with wide discretion to adopt frameworks to address the state's unique needs while complying with the provisions of the Convention.

In the past, individuals criticized UNCLOS and MARPOL because port states lacked regulatory powers in waters beyond two hundred miles from shore. Because "it is possible that an incident in those waters might pollute the coastal state itself or, more likely, its EEZ and continental shelf resources," powers extending beyond the EEZ could be crucial for a port state addressing a pollution problem. Despite this issue, the Ballast Water Convention does not expressly extend the port state's enforcement powers to the area outside of the port state's EEZ. Additionally, although the Ballast Water Convention does not include a provision that allows the flag state to supersede port state proceedings, the Ballast Water Convention indicates that "all possible efforts shall be made to avoid a ship being unduly detained or delayed" under the Convention, or the ship will be "entitled to compensation for any loss or damage suffered." Although it is unclear what the scope of the undue delay requirement is, the provision has the potential to undermine the port state's ability to board the vessel and enforce the provisions of the Convention.

Overall, the Convention, if ratified, represents a significant step forward in the attempt to address ballast water discharges and aquatic species introductions on an international scale. However, the Convention does not appear to go far enough, particularly without more stringent technology requirements, to force the adoption of effective treatment technologies. In the absence of a ratified treaty and the adoption of stringent technology requirements by individual countries, potentially within the Ballast Water Convention's structure, questions remain over what might force the adoption of effective treatment technologies. As we will next examine, although the United States has yet to adopt stringent performance standards for ballast water treatment technologies, developments at the national, regional, and state level may still have the power to trigger the shipping industry's adoption of effective treatment technologies, even in the absence of more stringent requirements at the international level.

The Management of Ballast Water Discharges in the U.S.

The United States has the power to provide leadership for the global community in the adoption of ballast water treatment technologies, but several challenges remain. The United States has yet to ratify the Ballast Water Convention or adopt comparable, aggressive domestic legislation with treatment technology performance standards. Most recently, in 2008, the Ninth Circuit held that EPA must regulate ballast water and other vessel discharges under the Clean Water Act (CWA). Despite the potential for adoption of performance standards under the CWA, EPA's ineffective response to the Ninth Circuit's holding has done little to advance the adoption of treatment technologies in the United States. Even in the absence of a national solution to ballast water discharges, however, this Part asserts that state and regional efforts may still have the power to drive the adoption of treatment technologies that are more stringent than those required under the Ballast Water Convention.

Federal Responses to Aquatic Invasive Species

Currently, the Nonindigenous Aquatic Nuisance Prevention and Control Act of 1990 (NANPCA), as amended by the National Invasive Species Act of 1996 (NISA), controls the Coast Guard's authority over ballast water discharges in the United States. Through NISA, which reauthorised and expanded NANPCA, Congress "provided ships entering U.S. ports from outside the [EEZ] a three-year window of opportunity to undertake a voluntary open-ocean exchange program with a mandatory reporting requirement." NISA also required that the Secretary of Transportation make the ballast water exchange requirements mandatory if the voluntary program is not "satisfactory." Under the Coast Guard's original regulations for implementing NISA, the Coast Guard required each ship holding ballast water to file a report twenty-four hours before entering U.S. waters and to have a ballast water management plan. During the "first 12 months of the program, only 12,170 of the 58,000 vessels arriving in U.S. ports filed a mandatory reporting form." Of the estimated 3500 vessels that declared an intention to discharge ballast water, only twenty-one percent of the vessels reported a full mid-ocean exchange of ballast water. Because the voluntary compliance with the ballast exchange requirements was too low, the Secretary imposed mandatory requirements for all incoming ships.

Current Coast Guard regulations require any ship with ballast water entering U.S. waters from outside the U.S. EEZ to either 1) conduct a mid-ocean ballast water exchange at least two hundred nautical miles from shore, in a location where the depth is at least two thousand metres, 2) retain the vessel's ballast water, or 3) utilise an alternative, environmentally sound method of ballast water control, as approved by the Coast Guard. The Coast Guard does not require vessels travelling only coastally, or within the two hundred mile EEZ, to conduct a mid-ocean ballast water exchange. The regulations also do not require vessels to shift their course solely for performing a ballast water exchange. Finally, the federal ballast regulations provide a safety exemption from the ballast water exchange requirement if weather or other limitations place the crew or vessel at risk.

The Coast Guard regulations require vessel operators arriving in U.S. ports or from locations outside the U.S. EEZ to submit ballast water reporting forms. The Coast Guard enforces ballast water management regulations and regularly boards vessels to check for compliance with the regulations. Additionally, if inspectors have reason to believe a vessel did not comply with the ballast water requirements, the Coast Guard can examine ballast water samples and assess criminal penalties or civil penalties of up to $27,500 per violation.

As described in Part H, ballast water exchange is an incomplete solution to the problems with invasive species; ballast water exchange, without the technology to destroy or remove pathogens and other aquatic species, is only partially effective. Additionally, ships often circumvent exchange requirements due to safety considerations. As such, "the current federal program has been criticized as inadequate," and individuals have criticized the Coast Guard for delays in adopting performance standards for ballast water discharges.

Although "there is wide agreement on the need for stronger measures to control ballast water discharges," there are differing views on how the government should implement controls. Over the last several years, differences in opinions stalled several proposed pieces of federal ballast water legislation. One of the bills that came closest to making it through Congress was House Resolution 2830, the Coast Guard Authorisation Act of 2007 (CGAA), which passed the House in April of 2008. The CGAA "would strengthen the existing provisions of law by amending NANPCA section 1101 to require that such vessels have a ballast water management plan, maintain record books, comply with ballast water exchange requirements, and comply with ballast water treatment requirements." Congress modelled the ballast water provisions of the CGAA after the provisions of the Ballast Water Convention, but imposed performance standards that are 100 times more stringent than the IMO standards; the CGAA, as passed by the House, limits discharged organisms to 0.1 living organisms greater than 50 micrometres in size per cubic metre, and to 0.1 living organisms per millilitre for organisms between 10 and 50 micrometres in dimension. Under the CGAA, the Coast Guard retains jurisdiction to develop regulations and monitor ballast water discharges.

Despite Congress' inclusion of more stringent performance standards than those of the Ballast Water Convention, political controversy remained over whether the bill should preempt more stringent state standards and whether EPA should retain regulatory power over ballast water discharges under the CWA. The potential preemption of state regulatory controls over ballast water, regardless of whether they are more or less stringent, is one of the primary reasons why states and others opposed the CGAA's approach. To resolve issues over federal preemption of state ballast water regulations, Senator Boxer circulated a compromise that would set either 1) California's ballast water standard, 2) any future CWA standard promulgated by EPA, or 3) a different standard set by the Secretary of Homeland Security, whichever is strictest, as the national standard when the technology requirement takes effect in 2012. This proposal would allow states to retain their current ballast water programs until 2012, and to participate in enforcement and technology certifications. Senator Boxer's draft also included a rigid 2012 deadline for ships to install treatment technologies. Despite Boxer's attempted compromise, H.R. 2830 did not pass before the adjournment of the 110th Congress in 2008.

Northwest Environmental Advocates

Although Congress has yet to pass comprehensive ballast water legislation or discharge performance standards, this Chapter asserts that the recent decision by the Ninth Circuit in Northwest Environmental Advocates has the potential to alter the regulatory scheme for ballast water discharges in the United States by bringing ballast water regulation under the purview of the CWA. This Part continues with an overview of the CWA's ability to act as a tool for regulating ballast water discharges, the Ninth Circuit's decision in Northwest Environmental Advocates, and EPA's response to the court's order.

The Clean Water Act as a Tool for Controlling Ballast Water Discharges

One debated issue is whether the permitting and standard-setting requirements of the CWA are the appropriate tools for managing ballast water discharges. Sections 301 and 402 of the CWA expressly prohibit any unpermitted "discharge of any pollutant" from a point source into navigable waters of the United States. The CWA defines "discharge of a pollutant" as "any addition of any pollutant to navigable waters from any point source." Congress further defined "point source" as "any ... vessel or other floating craft." The definition of pollutant expressly excludes "sewage from vessels" and "discharges incidental to the normal operation of a vessel of the Armed Forces," but includes "dredged spoil, solid waste, incinerator residue, sewage, garbage, sewage sludge, munitions, chemical wastes, biological materials, radioactive materials, heat, wrecked or discarded equipment, rock, sand, cellar dirt and industrial, municipal, and agricultural waste discharged into water." In 1973, EPA issued a regulation that exempted several categories of vessel discharges from the permitting requirements under the CWA. Under the exemption, EPA did not require a permit for "any discharge of sewage from vessels, effluent from properly functioning marine engines, laundry, shower, and galley sink wastes, or any other discharge incidental to the normal operation of a vessel." At the time, EPA justified the vessel exemption on the basis that "this type of discharge generally causes little pollution and exclusion of vessel wastes from the permit requirements will reduce administrative costs drastically."

The Ninth Circuit's Holding in Northwest Environmental Advocates

In 1999, a coalition of environmental organisations, including Northwest Environmental Advocates (collectively NWEA), petitioned EPA to repeal the vessel discharge exemption in 40 C.F.R. Section 122.3(a), claiming the CWA did not authorise the regulation. Ultimately, EPA denied NWEA's petition. NWEA subsequently filed a complaint in federal district court, asserting that EPA's refusal to rescind the regulation was "arbitrary, capricious, an abuse of discretion and otherwise not in accordance with the law ... and in excess of statutory jurisdiction, authority, or limitations." In 2005, the district court granted NWEA's motion for summary judgment and ordered EPA to repeal 40 C.F.R. [section] 122.3(a). During further proceedings in 2006 on the appropriate remedy, six states joined as plaintiffs and the Shipping Industry Ballast Water Coalition intervened as a defendant. As a result of the 2006 proceedings, the district court granted NWEA's motion for a permanent injunction, remanded the regulation to EPA, and set a deadline of September 30, 2008, for vacatur of the regulation. EPA and the Shipping Industry Ballast Water Coalition appealed to the Ninth Circuit Court of Appeals.

On appeal, the Ninth Circuit reviewed de novo the question of subject matter jurisdiction and the district court's grant of summary judgment. Because the decision involved a question over an agency's interpretation of a statute, the court applied the two-step inquiry articulated in Chevron, U.S.A., Inc. v. Natural Resources Defence Council, Inc. After concluding that the district court had subject matter jurisdiction and the statute of limitations did not bar NWEA's claim, the court turned to NWEA's ultra vires claim. EPA asserted three arguments. First, EPA argued that the court should only vacate 40 C.F.R. [section] 122.3(a) as it relates to ballast water discharges because NWEA did not discuss other discharges, such as gray water or discharges incidental to the normal operation of a vessel, in the 1999 petition to repeal the regulation. Second, EPA argued that either the CWA authorised the vessel exemption or, alternatively, the CWA is ambiguous on the subject and the court should defer to EPA's interpretation. Third, EPA asserted that Congress acquiesced to EPA's interpretation, even if the regulation was originally beyond the scope of the CWA.

In response to EPA's first argument, the court concluded NWEA challenged all three exemptions in the regulation, including gray water, ballast water, and discharges incidental to the normal operation of a vessel. Turning next to EPA's second argument, the Ninth Circuit considered whether the text of the CWA authorises EPA's exemption, or whether the court should defer to EPA's interpretation because of an ambiguity in the statute.

In reviewing the plain meaning of the text, the court determined that "Congress expressed 'a plain ... intent to require permits in any situation of pollution from point sources." The court reasoned that because section 402 uses the word "may," section 402 "allows the Administrator to issue a permit, but does not provide that the Administrator may entirely exempt certain categories of discharges from the permitting requirement." EPA did not "seriously contest" the court's holding on the second claim, and instead focused on the congressional acquiescence argument.

Next, the Ninth Circuit turned to EPA's third claim, which asserted that Congress subsequently acquiesced to EPA's regulation. Noting the high standard for finding congressional acquiescence, the Ninth Circuit analysed a number of statutes, including NISA and the NANPCA, in which Congress addressed the types of pollution covered in 40 C.F.R. [section] 122.3(a). However, the court concluded that the statutes did not contain the "overwhelming evidence" necessary to find congressional acquiescence. Accordingly, the court affirmed the decision of the district court and held that EPA acted beyond the scope of the agency's authority under the CWA in promulgating the regulatory exemption.

Significance of the Ninth Circuit Holding

The CWA's various provisions have the potential to be a powerful tool for controlling ballast water discharges. Because of the lack of binding performance standards for treatment technologies, the Ninth Circuit's holding in Northwest Environmental Advocates could have large repercussions for ballast water regulations in the United States.

First, under the provisions of the CWA, shippers visiting waters of the United States would need to obtain a permit under section 402 of the CWA.

Second, all discharges would need to comply with section 301, which requires the application of the best available technology that is economically achievable for vessels. The technology requirements could establish a uniform minimum level of treatment that is based on technology, which would go beyond ballast water exchange in terms of effectiveness. Finally, amongst other features, the CWA's criminal penalties and citizen suit provision would provide a strong enforcement mechanism to ensure that vessels comply with mandatory technology requirements.

Overall, the decision in Northwest Environmental Advocates could provide a trigger for the adoption of effective treatment technologies on a national scale by bringing ballast water discharges within the purview of the CWA. Further, to the extent that politicians, other individuals, or the shipping industry dislike the CWA as a mechanism for controlling ballast water discharges or are concerned about overlapping jurisdiction between EPA and the Coast Guard, the Ninth Circuit's holding could serve as the catalyst that drives Congress to adopt a statute that specifically addresses ballast water discharges and implements performance standards.

EPA's Response to Northwest Environmental Advocates

Following the Ninth Circuit's rifling in Northwest Environmental Advocates, EPA issued a national general permit to cover vessel discharges under the CWA. The ballast water provisions of the Vessel General Permit (VGP) are essentially a codification of existing Coast Guard regulations. The permit states "all discharges of ballast water may not contain oil, noxious liquid substances (NLSs), or hazardous substances in a manner prohibited by U.S. laws, including section 311 of the Clean Water Act." However, with regard to alien species, the VGP only requires that vessels "minimise or avoid uptake of ballast water" in areas known to have populations of harmful organisms or pathogens, and provides suggested control measures for minimising the spread of aquatic invasive species, such as the use of gray or potable water for ballast.

Similar to the Coast Guard regulations, the permit also requires that any vessels with ballast water taken on in areas less than two hundred nautical miles from shore, and that subsequently operate beyond the EEZ, must carry out an exchange of ballast water in an area more than two hundred nautical miles from shore. The permit also includes exemptions from the ballast exchange requirements; the master of the vessel can elect not to exchange the ballast water when necessary for safety reasons, when the ship uses other alternative ballast water management systems, or when the ship retains all ballast water for the duration of the vessel's voyage in U.S. waters. The permit does not require "any numeric treatment standards for the discharge of living organisms as part of this permit issuance and ... instead requires management practices (e.g. ballast water exchange) that decrease the risk of aquatic nuisance species introduction."

Several state regulators, environmental organisations, and others are dissatisfied with the terms of the VGP. For example, state officials from Oregon's Department of Environmental Quality told EPA "the permit 'perpetuates federal regulations that are insufficient for protecting Oregon's waterways.'" In certifying that the VGP complied with state water quality requirements pursuant to section 401 of the CWA, states such as California and New York, believing the VGP was not protective enough of state waters, "almost put a whole ballast program on top of the permit." In addition, industry and environmental groups filed lawsuits in several courts challenging EPA's general permit, with environmental groups claiming the permit is far too lenient and industry officials focusing on the extensive number of state requirements imposed on the permit through the state certification process.

Overall, EPA's response to Northwest Environmental Advocates has done little to change the regulatory framework already implemented by the Coast Guard in the United States. Although EPA acknowledges ballast water exchange helps reduce the risk of aquatic nuisance species, albeit with variable effectiveness, EPA chose not to impose any treatment requirements in the VGP. Without the imposition of mandatory treatment requirements, it is possible that the shipping industry will not spend the money to develop and implement the technologies necessary to control species introductions by ballast water, particularly since the shipping industry would run the risk of adopting technologies that might later be declared obsolete by the IMO or another governmental entity.

Nevertheless, there is hope; although the federal government has continued to delay implementation of a solution to drive technology, states—refusing to wait for federal action—have already set or have begun setting their own technology requirements for ballast water discharges.

Additionally, although it is unclear whether EPA will take action through settlement negotiations, litigation, or another administrative decision, EPA Administrator Lisa Jackson recently signalled that the agency might reconsider the permit, which "doesn't begin to address some of the concerns that are still out there." This Part continues with a discussion of how the ineffective VGP, as well as the lack of a federal statutory regime, could drive the adoption of treatment technologies that are more stringent than those required under the Ballast Water Convention.

State Action in the Absence of Federal Action

Motivated by delay in the development of international and domestic laws, several states created, or are in the process of developing, legislation to control ballast water discharges. Although California is on the forefront of implementing stringent discharge requirements, Oregon and Washington are close behind, and there is a high probability that the West Coast states, excluding Alaska, will implement a regional treatment standard if the federal government does not define a treatment standard in the near future. This Part discusses actions currently underway in various states, and argues that the state-driven regulation of ballast water could drive international adoption of effective treatment technologies for ballast water.

California on the Frontline

Responding to concerns over the lack of performance standards for ballast water discharges, the California State Land Commission (CSLC) developed and submitted ballast water standards to the California legislature in 2006. The California legislature incorporated the standards into the Coastal Ecosystems Protection Act of 2006 and passed the CSLC's recommendations into law. The CSLC based the performance standards on organism size and developed a phased implementation schedule based on vessel size class and construction date, which is similar to the IMO guidelines. California's standards require that there be no detectable riving organisms greater than fifty micrometres in dimension, and less than 0.01 living organisms per millilitre that are between ten and fifty micrometres in dimension. In other words, California's technology requirements overall are one thousand times more stringent than the Ballast Water Convention requirements and ten times more stringent than proposed requirements considered recently in Congress.

West Coast Alliances as a Catalyst for Change

Although Oregon's current regulatory program is similar to the Coast Guard's program, the state's Shipping Transport of Aquatic Invasive Species Task Force (Task Force) is encouraging the state to enact more stringent protections for Oregon waters in 2009. The purpose of the Task Force, which the legislature created in 2007, is to make recommendations for future amendments to Oregon's ballast management regulations. In recommendations to the 2009 Oregon Legislature, the Task Force indicated that "if sufficiently protective Federal standards are not in place by 2009," the state should pursue "an Oregon standard that is complementary to neighbouring states." Additionally, "if Washington and California adopt comparable treatment standards, the Oregon Department of Environmental Quality will adopt a common west coast treatment standard." In Washington, similar efforts to create a regional treatment standard are underway. The Washington Department of Fish and Wildlife (WDFW), in tandem with a Ballast Water Work Group, is rewriting Washington's laws to replace the state's current laws with concentration requirements that are comparable to California's laws. Additionally, "WDFW will no longer independently approve treatment systems for use in state waters and will instead rely on regional, national or international approvals."

Great Lakes Developments

The Great Lakes states have either enacted or are in the process of examining strict ballast water treatment requirements. For example, in Michigan, state law indicates that without a permit or another form of permission, "the discharge into the waters of the state from an oceangoing vessel of any ballast water is prima facie evidence" that a person discharged an injurious substance into Michigan's waters. Under the general permit that implements Michigan's ballast water law, if a ship wants to discharge ballast water, the vessel must use approved, "environmentally sound and effective technology" for "preventing the discharge of aquatic nuisance species" and must ensure aquatic species are not discharged "at levels which are injurious to the designated uses of the waters of the state." In other words, Michigan's standards do not allow for the discharge of any aquatic invasive species that could harm the environment. Minnesota followed Michigan's lead and adopted ballast water regulations, but Minnesota's requirements are less stringent than those in California or Michigan; under Minnesota's vessel general permit, ships must meet treatment requirements that are comparable to those under the Ballast Water Convention. Although none of the other Great Lakes states have permitting regimes as of yet, several states in the region are examining options for ballast water legislation.

In New York, recent developments related to the state's certification of EPA's VGP illustrate how the state is taking action in the absence of effective federal controls for ballast water. Relying on the state's narrative water quality standards and antidegradation policy, New York imposed performance standards and an implementation schedule for treatment technologies as part of the VGP certification. Under New York's conditions, with certain limited exceptions, each vessel, by no later than January 1, 2012, must have a ballast water treatment system that discharges less than one living organism that is fifty or more micrometres in dimension per ten cubic metres and less than one living organism that is less than fifty micrometres in dimension per ten millilitres. Despite the fact that New York did not have a regulatory or statutory version of the certification's performance standards, a New York Supreme Court upheld the conditions as "rationally derived from the authority of the New York Department of Environmental Quality to control ballast water pollution." Ultimately, whether other states follow Michigan, Minnesota, and New York's lead in the Great Lakes region, and whether the states will work toward regional uniformity has yet to be seen. However, as demonstrated by the Great Lakes states thus far, the states are concerned about aquatic invasive species introductions and are likely to continue working towards the implementation of ballast water controls.

State Attempts to Spur the Adoption of Treatment Technologies

There is significant support for the proposition that environmental regulations can induce industries to adopt environmental technologies. For example, under the Clean Air Act (CAA), Congress granted California the authority to regulate motor vehicle emissions and preempted the vehicle emission standards of all other states. Although there were concerns that California's exemption "would undermine the efficiency of uniform federal product standards by imposing a parallel set of conflicting standards on industry," the "concerns have proven to be unwarranted." California, through its CAA exemption, led the way in creating rigorous standards and in promoting the development and adoption of new technologies. Ultimately, California's example in the CAA context demonstrates that states can lead the way in adopting standards and promoting new technologies.

With California, Oregon, and Washington moving toward regional uniformity with ballast water regulations, and with developments in the Great Lakes region, state action may have the power to trigger the adoption of treatment technologies, even in the absence of binding federal or international ballast water controls. In total, over one third of the value of U.S. waterborne foreign trade comes through West Coast ports and another 16% comes through the states bordering the Great Lakes, which totals to over half the value of U.S. waterborne foreign trade.

Additionally, in terms of sheer volume, approximately 30% of the total number of metric tons imported and exported from the United States goes through the West Coast and another 14% through the Great Lakes region. As a rough estimate of the number of vessels that enter West Coast waters, California's arrival statistics provide some insight into the number of vessels that enter West Coast waters and would need to meet the stringent performance standards. For example, between 2000 and 2008, 5682 unique vessels with ballast water capacities of greater than 5000 metric tons entered California's ports.

Just as California was able to promote more rigorous vehicle emissions technologies through the adoption of Stringent standards, the regional coalition of California, Oregon, and Washington may likewise be able to drive the adoption of technologies that are more stringent than those required under the Ballast Water Convention. With Great Lakes states moving towards the adoption of performance standards, and potentially towards regional cooperation or uniformity in the future, the Great Lakes region may also have the power to drive the adoption of treatment technologies. Although the shipping industry might have concerns over the potential for several countries and states to issue regulations independently, the fact that California's treatment standards are more stringent than the Regulation D-2 requirements under the IMO should not deter the shipping industry since the standards in California, Minnesota, Michigan, and New York are equal to or go beyond the Ballast Water Convention's baseline performance standards. Further, since the Ballast Water Convention allows countries to enact more stringent performance standards, a West Coast or Great Lakes standard is unlikely to create complications that go beyond the complications already likely to be present from variations allowed under the Ballast Water Convention and would sync the requirements for a large area of the United States, thereby reducing at least some difficulties for the shipping industry.

Another legitimate concern by the shipping industry might be the potential conflicts between technologies approved by the IMO and technologies that meet the discharge standards in California or other states. Because the IMO must approve treatment technologies to comply with the terms of the Ballast Water Convention, vessel owners anxious to meet California's treatment requirements run the risk of adopting technologies that will ultimately be disapproved by the IMO. However, as illustrated by Washington, states recognise the potential conflicts between technology approval processes and are working to reduce the barriers to the adoption of treatment technologies by allowing vessels to utilise technologies approved through regional, national, or international mechanisms—so long as the technologies meet the states' performance standards, which exceed and are compatible with the Ballast Water Convention. Ultimately, the state provisions could force the adoption of treatment technologies that meet the stringent state performance standards, but also comply with the IMO's technology approval procedures.

Looking Toward a Brighter Future for Ballast Water Controls

With the increasing number of problems arising from ballast water discharges and aquatic invasive species introductions, the United States will need to decide how to best implement performance standards for ballast water. This Part analyses the various options for the United States in adopting a comprehensive ballast water framework and offers suggestions on how the United States might reconcile the differences between federal, state, and international regulatory options.

Choosing Between Policy Options

One of the largest, looming issues that the United States will need to address in the near future is how to regulate ballast water discharges. The United States has three main options for implementing ballast water controls: 1) ratify the Ballast Water Convention, 2) allow states to develop and impose performance standards under the CWA or their own regulatory mechanism, or 3) utilise a federal program for implementing performance standards, whether under the CWA or a standalone program.

With regard to whether the United States should adopt the Ballast Water Convention over other options, there are a number of policy issues the United States should consider. The Ballast Water Convention would provide an international floor for regulation of ballast water discharges, and the United States' ratification of the treaty would bolster current international efforts to implement treatment technologies. The United States should also keep in mind a number of conflicting tensions between state, federal, and international regulatory mechanisms, which could impact the effectiveness of the adopted framework. First, as described earlier, the United States will need to adopt significantly more stringent treatment requirements than those contained in Regulation D-2 to ensure the protection of the country's coastal waters since the Ballast Water Convention's treatment standard is potentially too low to eliminate species introductions. Second, the Ballast Water Convention's timelines are potentially too relaxed and should be tightened if the United States hopes to spur the adoption of treatment technologies in the near future. Additionally, under the broad language of Article 4 of the Ballast Water Convention, the United States may be able to utilise the provisions in the Ballast Water Convention to adopt national policies that take into consideration the "particular conditions" of the United States and harmonize the domestic and international regulatory frameworks. With state controls for ballast water discharges, one concern voiced is that "an increasing number of state-by-state regulations can lead to a complex patchwork of potentially disparate regulations, leading to inherent challenges in consistency, enforceability, and workability in foreign trade." However, as shown by California, Oregon, and Washington, states are working to implement uniform standards at the regional level, which will alleviate the burdens of trans-jurisdictional regulation on the shipping industry. Additionally, although Congress has yet to pass a federal statute setting mandatory treatment requirements for ballast water discharges, the current efforts for compromise in the U.S. Senate indicate that California's stringent standards may ultimately have the power to drive the selection of the federal standards.

Third, there are also numerous policy benefits to implementing controls under the CWA. One reason why advocates prefer the CWA to other regulatory mechanisms is that the statute allows states to adopt more stringent requirements than those required under the CWA. Although the preservation of state sovereign power could lead to the adoption of more stringent treatment requirements, there is arguably a large difference between the regulation of stationary sources and the regulation of the shipping industry. Unlike industries typically regulated under the CWA, which are fixed in a particular location, the movement of vessels across various jurisdictions presents a unique challenge for regulating discharges since vessels would need to comply with variations in discharge requirements as they move from port to port. Because the shipping industry travels across international lines and waters, the shipping industry is likely to have difficulty in meeting a wide range of individualised, state-by-state treatment requirements. Conversely, as California demonstrated under the CAA, federal preemption may not be the best solution; by giving states a limited amount of regulatory authority, as under the CWA, where states can adopt more stringent, but not less stringent, standards, the regulatory mechanism could promote and lead to the adoption of more rigorous technologies and standards. Although state controls may sometimes be more effective than federal controls at spurring the development of new technologies, as shown by California under the CAA, there are also likely to be additional costs and complexities if the shipping industry needs to comply with performance standards that vary from state to state. Thus, it will be important for Congress to address whether federal preemption is the best solution for dealing with ballast water problems.

Another policy choice related to the CWA is over which agency should have jurisdiction to administer vessel discharge permitting and monitoring. Some critics argue that it would be preferable for EPA to control vessel discharges "because EPA's primary mission is to protect public health and welfare," whereas the Coast Guard "has shown insufficient interest in pollution control generally, and ballast water management specifically." Additionally, overlapping jurisdiction between the two agencies is likely to lead to problems, as "overlapping regulation can lead to confusion, high compliance costs, and a drag on otherwise beneficial activities."

Finally, in addition to dealing with the technology-based performance standards under the CWA, Congress also faces the challenge of whether to preempt state water quality standards, which are distinct from performance standards under the CWA. As illustrated by New York's certification of the VGP, which relied upon state water quality standards when imposing technology-based performance standards, water quality standards are an important state check on potentially unprotective performance standards. Section 303 of the CWA requires states to develop water quality standards for waters within the states' borders. For state water quality standards, the state most first define "the water quality goals of a water body, or portion thereof, by designating the use or uses." Second, the state must determine the level of water quality needed to protect the designated uses and develop criteria for protecting the designated uses. Thus, water quality standards set a "floor" for water quality and EPA must alter the technology-based performance standards to achieve the state's water quality standards. Although water quality standards can play an important role for states in protecting their waters, because the imposition of the criteria could lead to diverse, state-by-state inconsistencies in ballast water controls, Congress will need to decide whether states should retain their ability to impose additional treatment requirements for ballast water using state water quality standards.

A New Federal Framework

Although the three main options for controlling ballast water discharges in the United States—ratifying the Ballast Water Convention, continuing state-led programs, or pursuing a federal program—appear distinct, there are ways to potentially harmonize the seemingly conflicting policies and laws behind each option. In light of the current debates in Congress over the Ballast Water Convention, it appears unlikely that the United States would comply with the provisions of the Convention verbatim. As such, in the absence of a ballast water bill, it appears likely that the United States could end up with a framework that either implements ballast water controls under the CWA or through an alternate program. This Chapter suggests that the appropriate federal framework should build upon the CWA as the vehicle for ballast water controls, but enact certain provisions that adjust the CWA to account for differences between the shipping industry and other point sources regulated under the CWA.

First, the United States needs to implement performance standards that dictate the number of viable organisms allowed per litre or millilitre of ballast water. Implementing an effective standard for the technology requirements is essential to ensuring the elimination of viable species from ballast water and the elimination of aquatic invasive species introductions from waters of the United States. Additionally, the United States, regardless of whether it ratifies the Ballast Water Convention, should implement a system that provides for use of treatment technologies approved by the IMO. Although the United States could retain the power to reject unsafe or unacceptable treatment technologies, the United States should move towards uniformity with the IMO in the technology approval process to ensure vessels plying U.S. waters do not run the risk of adopting technologies that are later rejected under the IMO framework. By providing a degree of certainty for vessels and by reducing the risk that they will install the "wrong" technology, the United States might increase compliance amongst vessels.

Second, because vessels are mobile, and therefore subject to changing jurisdictional requirements, the United States should ensure the performance standard will create greater consistency and predictability for the shipping industry to comply with regulations. Thus, the federal government should allow states to adopt more stringent performance standards, but allow this decision to take place within the confines of a model similar to that employed under the CAA. In general, the CAA preempts state standards "relating to the control of emissions from new motor vehicles and new motor engines." Since California had significant problems with mobile source pollution and was the only state with vehicle emissions standards in place prior to the enactment of the CAA, Congress provided an exemption from the prohibition for California, provided the state meets certain criteria. EPA, following notice and comment, must grant a waiver for California if 1) the state determines that the "standards will be, in the aggregate, at least as protective of public health and welfare as applicable federal standards" and 2) EPA does not find that California's determination is arbitrary and capricious, that California does not need the standards "to meet compelling and extraordinary conditions," or that California's standards and enforcement procedures are inconsistent with section 202 of the CAA. Additionally, Congress allowed other states to adopt California's more stringent standards subject to certain conditions, including a requirement that states must utilise standards that are identical to California standards and for which EPA issued a waiver.

Using the CAA as a model, the federal government could enact a provision that sets California's performance standards, the most stringent in the nation, as an alternate to the federal regime. The system could also utilise the waiver provisions of the CAA in the CWA context; the system would allow states interested in adopting more stringent performance standards to apply for a waiver and justify their need for the higher standard. EPA could then consider the merits of the application and issue a waiver to the state, thereby allowing the state to adopt the more stringent performance standard. Ultimately, if the federal government also implements national water quality standards, as described below, the United States could wind up with only two different standards—the federal standard and the more stringent state standard. Although this would not create total uniformity for the shipping industry, the solution could minimise the industry's difficulty in dealing with multiple states, each with a unique regulatory system, while still allowing states to be protective of their waters. Overall, the process would create national uniformity and a predictable system for adjusting the standards, while providing a mechanism for creating greater regulatory certainty. The public process involved in the waiver procedures would also place the shipping industry on notice about future changes in state standards.

Third, Congress will need to address whether to preempt the states' authority to develop and impose water quality standards. Unlike the CAA, where the statute establishes national air quality standards, the CWA requires states to develop and implement state water quality standards. Although technology-based performance standards play a vital role in spurring the adoption of treatment technologies, Congress must also address water quality standards since state water quality standards have the potential to undermine the consistency and predictability of the national framework; in other words, although California might have the power to drive the technology-based standards under the CWA, without national water quality standards, states would still be able to impose more stringent requirements through state water quality standards, thereby exceeding California's standards and creating inconsistencies in the regulatory framework.

Because variations in state water quality standards could lead to continued inconsistencies in the implementation of ballast water controls, and would do little to remedy the problems faced by the shipping industry in complying with a wide variety of state-by-state frameworks, Congress should preempt individualised state water quality criteria and implement a national water quality criteria system akin to that under the CAA. In determining a uniform water quality standard, the United States should model the provision on the most stringent state water quality standard in the United States as the national standard. Adopting the most protective state water quality standard in the nation would ensure that states are able to protect the quality of their waters, would provide greater consistency on a national scale, and could continue to drive the adoption of treatment technologies. Overall, a uniform, but stringent, set of national water quality standards could balance the needs of vessels with the desire of states to protect the quality of their waters.

Fourth, the United States should ratify the Ballast Water Convention and integrate the Convention's requirements into the CWA. As illustrated by the integration of the Montreal Protocol into Title VI of the CAA, the United States could ratify and integrate provisions of the Ballast Water Convention into existing U.S. laws. Under Title VI of the CAA, Congress codified many of the goals of the Montreal Protocol and, in several areas, went beyond the requirements of the Protocol. Similar to Article 2 of the Ballast Water Convention, which allows states to adopt more stringent standards to prevent, minimise, and eliminate the transfer of invasive species, the Montreal Protocol contains a provision allowing states to adopt more stringent performance standards than those required under the treaty.

Thus, in integrating the Montreal Protocol into the CAA, Congress was able to make a number of alterations to the international regime, while remaining within the confines of the Protocol. For example, Title VI contained a more aggressive schedule for phasing out harmful substances, imposed recycling requirements that were not in the Protocol, and adopted a program for phasing out less harmful, but still destructive chemicals, while the Protocol did not include a provision dealing with less harmful substances. Title VI also includes several provisions to facilitate the United States' compliance with the Protocol. For example, Congress directed EPA to fully implement the United States' obligations under the Protocol and instructed the President to take actions such as banning the export of technologies used to produce substances harmful to the ozone layer.

The ability of the United States to integrate the Montreal Protocol into the CAA exemplifies how the United States could integrate the provisions of the Ballast Water Convention into the CWA. Relying on the provisions in the Ballast Water Convention that preserve the United States' right to enact more stringent ballast water measures, the United States could ratify the Ballast Water Convention, enact dual performance standards that follow the CAA's framework, and utilise the provisions in Article 4 that allow the United States to adopt national policies for achieving the objectives of the Ballast Water Convention. Utilising Articles 2 and 4 of the Ballast Water Convention, the

United States could expressly adopt the Convention, but also ensure the protection of U.S. waters by enacting performance standards and other measures that go beyond the international standard. Ultimately, if the United States were to integrate the Ballast Water Convention into the CWA, the United States could demonstrate critical leadership on a serious environmental threat, protect the waters of the United States, and spur the industry to adopt effective treatment technologies.

Ballast water discharges are one of the primary vectors for inadvertent introductions of invasive species into aquatic ecosystems, yet to date there are few binding, effective legal regimes in place to prevent species introductions or to force the adoption of treatment technologies. At the federal and international levels, there is growing support for the IMO's Ballast Water Convention. Despite the fact that the Ballast Water Convention is not yet in force, countries and states, such as California, are already modelling binding legislation after provisions of the Convention, thereby increasing the efficacy of international regulation of ballast water.

Although the Convention may not be as stringent as desired by the United States, the United States should ratify the convention in light of the country's ability to retain its sovereign power to enact more stringent legislation domestically. Ratification would bring the world one step closer to achieving international controls for reducing the impact of aquatic invasive species and would drive adoption of effective ballast water controls, while allowing the United States to adopt stringent performance and water quality standards. Nevertheless, in the absence of federal or international actions to implement ballast water controls, the adoption of stringent performance standards in California and across the West coast may still have the potential to drive the adoption of treatment technologies. There is hope, however, that in the wake of state actions and the decision in Northwest Environmental Advocates, the United States will enact a comprehensive ballast water framework and work towards halting the ballast water hitchhikers.

The Definition of "Discharge": Section 401 of the Clean Water Act

On September 27, 1996, Judge Ancer Haggerty of the U.S. District Court for the District of Oregon handed down an opinion that has since caused controversy and debate among the environmentalists, ranchers, loggers, government agencies, and citizens who are affected by its far-reaching holding. In Oregon Natural Desert Association (ONDA) v. Thomas, the court held that the United States Forest Service had violated section 401 of the Federal Water Pollution Control Act (Clean Water Act (CWA)) by issuing a permit for grazing in the Malheur National Forest without first obtaining certification from the State of Oregon that the grazing activity would not violate Oregon's water quality standards, Section 401 requires states to certify that any federally permitted activity "which may result in any discharge" will not violate the water quality standards of the state in which the discharge will occur. In ONDA, the court reasoned that the term "discharge," as it is used in section 401, is not limited to discharges from point sources, Support for this reasoning was found in the plain meaning of "discharge," the statutory definition of "discharge," and the legislative history of the CWA. The ONDA decision is the first in the nation to hold that section 401 of the CWA applies to "all federally permitted activities that may result in a discharge, including discharges from nonpoint sources," and that pollution caused by cattle grazing constitutes a discharge within the meaning of section 401 of the CWA. If this broad interpretation of the term "discharge" is upheld, the implications will be far-reaching. Not only would every grazing permit issued by the Forest Service or the Bureau of Land Management (BLM) be subject to section 401 of the CWA, but so might permits for logging roads, mining activities, and any other nonpoint source activity potentially resulting in a discharge.

From a water quality standpoint, this strengthened state control over federal land use could provide our tired public rangeland, rivers, and forests with a much needed respite from the exhausting multitude of uses that currently degrade water quality. The United States Environmental Protection Agency (EPA) consistently reports that nonpoint source pollution, such as livestock grazing, is the leading source of degradation of water quality in the United States. By giving states the opportunity to regulate federally permitted activities within their borders to ensure compliance with water quality standards, section 401 may offer a partial solution to the complex nonpoint source pollution problem. There can be little doubt that requiring federally permitted activities to meet state water quality standards would do wonders for water quality, the ecological integrity of riparian zones and freshwater ecosystems, and the fish and wildlife species that depend upon those systems.

However, the ONDA decision has done more to open up debate on the proper scope and application of section 401 of the CWA than it has to resolve the issue. The intervening industry defendants were quick to appeal the decision to the Ninth Circuit Court of Appeals, with their position strengthened by an Idaho district court opinion holding that section 401 of the CWA did not apply to a permit for construction of a logging road, a nonpoint source, on national forest land. In Idaho Conservation League v. Caswell, the court reasoned that statutory definitions and the regulatory scheme of the Act clearly limit the term discharge in section 401 to mean point source discharges only. The federal government's reaction to the ONDA holding can be fairly characterised as scattered and divided. In January of 1996, the Department of Justice (DO J) filed a "protective notice of appeal," buying time to formally decide whether or not to appeal the case on behalf of the Forest Service.

In February of 1996, EPA made the surprising move of drafting revised section 401 implementation regulations which made clear that according to EPA, "discharge includes nonpoint sources." EPA apparently was criticized by other agencies for moving forward with a regulation before the administration had decided whether or not to appeal, and was forced to table the rulemaking pending conclusion of the litigation. Finally, on June 16, 1997, the Clinton Administration decided to appeal the ONDA decision despite opposition from EPA, the regulatory agency charged with implementing and enforcing the CWA.

The government's appeal marked the end of months of debate between EPA, the Forest Service, and BLM, and the end of a debate which EPA clearly lost. Thus, the Ninth Circuit Court of Appeals faces the task of determining how Congress intended section 401 of the CWA to be interpreted and applied. This Comment addresses the question of whether the term "discharge," as it is used in section 401 of the CWA, includes nonpoint sources of pollution. It offers a background sketch of the provisions of the CWA relevant to a discussion of section 401; this section is intended for readers who are unfamiliar with the CWA. It identifies the complex ecological impacts that federally permitted nonpoint source activities have on water quality. Part IV explains how section 401 of the CWA functions, and how it was used by states prior to the ONDA decision. Part V uses traditional tools of statutory construction to discern the intent of Congress in drafting section 401.

By analysing the plain meaning of discharge and its use in the CWA, the context of section 401 in the CWA regulatory scheme, and the legislative history of section 401, this section sets out the arguments in support of both broad and narrow readings of the term discharge in section 401. This concludes that while the CWA is ambiguous and contradictory in many respects, the plain meaning and legislative history of section 401 generally support a broad reading of the term discharge that includes discharges from nonpoint sources. It attempts to evaluate the effectiveness of section 401 state certification of nonpoint source activities as a water pollution control mechanism, in part by using Oregon's response to the ONDA decision as a case study.

Water Pollution Control: A Historical Perspective

To best understand the current regulatory scheme used to address the problem of water pollution in the United States, it is helpful to first review the historical context in which it evolved. The earliest federal legislation drafted in direct response to public concern about water quality was the Water Quality Act of 1948, which provided funding for the research and implementation of state water pollution control programs. As industrial development caused increasingly more serious water pollution problems, Congress provided additional funding to the states pursuant to the Federal Water Pollution

Control Act of 1956. Again in 1965, Congress was compelled to strengthen its directive to the states, this time requiring states to adopt water quality standards and implementation plans. However, this Act still did not provide for effective federal enforcement mechanisms, and by 1972 only half the states had set water quality standards. Further more, even when a state had adopted water quality standards, the standards were not effectively translated into requirements for individual dischargers.

In 1971, the Committee on Public Works (chaired by Senator Edmund Muskie) concluded that "the national effort to abate and control water pollution is inadequate in every vital aspect." The Committee perceived the core of the problem to be the inherent difficulty of linking the violation of a water quality standard to a particular pollutant, polluter, or water use. It was time for a dramatically different approach to water pollution control, and the solution to the problem was simple. Water pollution needed to be regulated at its source, in addition to being measured against water quality standards. It was in the context of this realisation that the regulatory universe of the Clean Water Act (CWA) was conceived.

Overview of the Clean Water Act

The overriding purpose of the CWA is to "restore and maintain the chemical, physical, and biological integrity of the Nation's waters." The Act seeks to achieve these goals by combining state water quality standards with the technology-based approach of setting effluent limitations on what a point source can discharge into the waters.

Definitions of "Point" & "Nonpoint" Source in the CWA

For regulatory purposes, the Act classifies water pollution into one of two categories, based on the source of the pollution. A "point source" is defined in the CWA as "any discernible, confined and discrete conveyance, including but not limited to any pipe, ditch, channel, tunnel, conduit, well, discrete fissure, container, rolling stock, concentrated animal feeding operation, vessel or other floating craft, from which pollutants are or may be discharged." Under this definition of point source, even a discrete conveyance is not a point source if it does not discharge, or potentially discharge, one of the "pollutants" specified by the CWA. Discharge of a pollutant from a point source is prohibited by the CWA, unless the discharger has a National Pollutant Discharge Elimination System (NPDES) permit. The end-of-the-pipe effluent limitations required by these permits contain both technology-based and water quality-based limits, and are at the heart of the CWA regulation of point sources.

The term "nonpoint source" is not expressly defined in the CWA, although it appears in several provisions, EPA defines nonpoint source pollution as pollution "caused by diffuse sources that are not regulated as point sources and normally... associated with agricultural, silvicultural and urban runoff, runoff from construction activities, etc." A nonpoint source can be understood as any source of pollution that cannot be traced to a point source. However, within this vast universe known as nonpoint source pollution, there is a wide variety of human-induced pollution-causing activities. The CWA drafters acknowledged the many causes of pollution by defining it broadly in the CWA as "the man-made or man-induced alteration of the chemical, physical, biological, and radiological integrity of water." A sampling of sources of nonpoint source pollution can illustrate the spectrum these sources represent. For example, pipes that discharge irrigation return flows are nonpoint sources because they are specifically exempted from the point source provisions of the Act. Courts have treated dams that discharge water with a low dissolved oxygen content as nonpoint sources because they do not discharge a "pollutant." Runoff from silviculture, agriculture, mining, or construction is nonpoint source pollution, unless is has been channelled by human efforts, and thus transformed into a point source of pollution. Nonpoint source pollution also includes erosion, addition of materials not formally considered "pollutants," and changes in water conditions. Changes in water conditions such as pH, temperature, or oxygen content are usually caused by activities lumped into the nonpoint source category, although "heat" is specifically listed as a "pollutant" in the CWA definition.

The Significance of State Water Quality Standards in the CWA Scheme

Regulation of point sources through the imposition of effluent limitations represented a new and necessary approach to water pollution control, but Congress recognised that water quality standards would remain a vital approach to regulation of both point and nonpoint sources of pollution. The 1972 CWA required that each state set water quality standards to be used in conjunction with technology-based limitations to achieve the goals of the Act. State water quality standards must be specific to particular bodies of water, and they consist of two basic elements. First, states must assign "designated uses" that establish the uses for which the waters must be protected. Designated uses must include, among other uses, propagation of fish, shellfish, and wildlife, as well as recreation in and on the waters. Second, the states must set water quality criteria that define levels of toxic pollution at which the designated uses will be compromised, and above which the body of water will violate water quality standards. Criteria can be in the form of a numeric standard or a broad narrative phrase, thereby providing a substantive structure for protecting designated uses. Water quality standards are an enormously important regulatory tool, because they give states the power to achieve a degree of water quality that could not be achieved with economically based effluent limitations alone.

Regulation of Nonpoint Sources under the CWA

Congress did not address regulation of nonpoint source pollution with a specific statutory provision until 1987, when it amended the CWA to include section 208 requirements for "areawide waste treatment management." Current regulation of nonpoint sources under the CWA is the subject of much discourse and research, mostly because the CWA provisions attempting to regulate nonpoint source pollution have proven largely ineffective. Sections 208 and 319 of the CWA are the two primary federal regulatory devices for controlling the problem of nonpoint source pollution.

Section 208 of the CWA was included in the original 1972 Act. It requires each state to 1) "identify... agriculturally and silviculturally related nonpoint sources of pollution... and their cumulative effects, runoff from manure disposal areas, and from land used for livestock and crop production" and 2) "set forth procedures and methods (including land use requirements) to control to the extent feasible such sources." In response, agencies looked at ways to modify land uses causing nonpoint source pollution, and created "Best Management Practices" (BMPs) that would prevent the pollution from occurring. BMPs range anywhere from treatment requirements, to the prohibition of certain practices, to maintenance procedures. BMPs have been compared to effluent limitations because they attempt to control nonpoint source pollution at its source and with a technology-based approach. In theory, BMPs could be an effective way to meet desired water quality standards, but in reality they have fallen far short of their goals. The failure of section 208 to reduce nonpoint source pollution has been attributed to a combination of lack of federal funding, inadequate water quality data, political conflict, and EPA's poor leadership.

In 1987, Congress amended the CWA to address the persistent problem of nonpoint source pollution. Section 319 basically makes states more accountable to EPA for the success of their section 208 plans by requiring states to report to EPA their success in meeting implementation schedules and by threatening funding cutoffs if the states fail to make satisfactory progress. While it is not clear whether these programs are faulty conceptually or just in their implementation, the fact remains that water quality reports reflect that nonpoint source pollution runs unabated into our nation's waters.

Nonpoint Source Pollution on Public Lands

Nonpoint Source Pollution Generally

Despite more than two decades of attempts to curb nonpoint source pollution, this pollution continues to have an enormous impact on our nation's waters. According to the most recent Environmental Protection Agency study, states report that water quality standards are not met in roughly 36% of their waters, and polluted runoff is the major cause cited for that failure. EPA reports that the leading pollutants impairing our nation's waters are bacteria, siltation, nutrients, oxygen-depleting substances, metals, habitat alterations, and suspended solids. Each of these pollutants plays a complicated role in aquatic ecosystems, and often the result of those ecological dynamics is water quality that cannot support the uses we have designated for protection (recreation, fishing, wildlife, etc.) For example, when riparian vegetation is damaged, a stream will often experience increased erosion and sediment load, reduced beneficial filtering of runoff pollutants, and higher temperatures due to less shade. Whether these reactions are "pollutants" or "pollution," they are likely to violate state water quality standards. High stream temperatures decrease the dissolved oxygen present in the water, and the fish and other organisms that depend on the oxygen cannot survive in the stream.

At the heart of the connection between nonpoint source pollution and compromised water quality is a truth articulated by EPA: "The health of rivers and streams is directly linked to habitat integrity on shore." Many of the nonpoint sources that cause serious degradation of water quality actually do so by degrading the ecological integrity of the upland or riparian environment in a way that leads to siltation, increased nutrients, habitat modification, or thermal pollution. How should these complex ecological reactions that result in water pollution be regulated? If section 401 does apply to nonpoint source pollution, does it apply to all federally permitted activities that cause water pollution, or just to activities that cause pollution as a result of a discharge? Consider these questions in the context of nonpoint source pollution on public lands.

Specific Nonpoint Sources of Pollution on Public Lands

The causes and effects of nonpoint source pollution are especially prevalent on public lands, where many activities that cause nonpoint source pollution are daily occurrences, such as livestock grazing, timber cuts and roads, and resource extraction. Public lands comprise 29% of the nation's land, and nearly half of the land in the "eleven western states." Furthermore, "almost two-thirds of the run-off in the eleven western states, and all the great western rivers, originate on the public lands." For these reasons, the land use activities that take place on our public lands have far-reaching effects on water quality.

Water Quality Effects of Livestock Grazing on Public Lands

The story of private livestock grazing on our public lands is a true "tragedy of the commons," and the subject of much current research as well as political controversy. Historical facts can help to illustrate the dramatic ecological transformation the western landscape has undergone as a result of livestock grazing: in 1870, Arizona Territory held about 5,000 cattle; twenty-one years later there were 1.5 million in the Arizona Territory and 26.5 million total in the western United States. Clean Water Act scholar Oliver Houck recently observed that ecosystem management of the public range faces "two great challenges... that the ecosystem we see today has been so greatly altered... and that the major source of the alteration, large numbers of cattle, remains." Domestic livestock grazing is a predominant use of much federal land, second only to recreation. Not surprisingly, the ecological impacts of grazing are not limited to the land. EPA reports that agricultural runoff contaminates 60% of the impaired river miles in the United States, and that livestock grazing is the second major source of that pollution after crop production.

Livestock grazing impacts on water quality result from the tendency of cattle to graze and remain in riparian zones; they consume all available vegetation and young woody plants, trample the banks, and defecate in the water. One commentator estimates that "at least 1000 tons of bovine excrement are discharged into our public waters every day." Livestock manure introduces bacteria into water, as well as excess nutrients, which in turn lower the amounts of dissolved oxygen available for fish and other species. Grazing in riparian zones loosens soil and eliminates riparian vegetation that would normally prevent erosion; studies have revealed that increases in water-borne fine sediments result in decreases of 37%-59% in biological productivity of a water body. Habitat modification is perhaps the most obvious water quality impact from livestock grazing; the combination of siltation and high temperatures impairs the habitat of trout and other salmonids. One commentator also speculates that the direct results of grazing (erosion and depleted riparian vegetation) cause other runoff pollutants to be more serious problems than they would be if they were filtered by a healthy riparian and aquatic ecosystem. The negative impacts of livestock grazing on water quality are mainly a result of discharges of manure and increased sediment from trampling and removal of riparian vegetation.

Water Quality Impacts of Timber Harvest on Public Lands

Timber production has long been recognised as a nonpoint source of water pollution, and much of our nation's timber production takes place on public lands. The silviculture activities that produce pollutants are related either to road building, transportation access, or actual harvesting of the timber. Increased sediment, or siltation, is the most common form of pollution resulting from silvicultural activities. In the past decade, there is an increasing body of evidence showing that clearcutting dramatically increases natural rates of landslide erosion. For example, one district court examining the impacts of Forest Service actions on anadromous fish populations in the Siuslaw National Forest found that a vast majority of the landslides in the forest were a result of clearcutting and timber roads. The court found evidence that those landslides caused debris torrents that "scour a stream channel to bedrock, damage streambank vegetation, and undercut adjacent slopes... removing not only fish spawning gravel, but also rearing grounds" and blocking fish passages.

Excessive nutrients are introduced by fertilizer and direct runoff of slash and debris, and this added organic material consumes oxygen during decomposition. Additionally, clear cuts and removal of canopy cover causes an increase in water temperature.

This thermal pollution, like that caused by livestock grazing, can increase bacteria levels and destroy fish habitat. Recognising the impact of silviculture on fish habitat, timber harvests have been completely prohibited in the Riparian Habitat Conservation buffer zones established by the Forest Service and the Bureau of Land Management to ensure the conservation of threatened and endangered salmon species. Water quality violations resulting from timber removal operations on public lands are caused by both clear discharge events and fuzzier non-discharge pollution like riparian vegetation removal; but in all cases, they are a serious problem for humans, fish, and wildlife.

Water Quality Impacts from Mining and Other Activities on Public Lands

A significant percentage of thc resource extraction that takes place in the United States occurs on public lands. While some mining processes are subject to regulation as point sources under the CWA (discharges from pipes, impoundments, sumps, ditches etc.), other mining activities affect water quality due to nonpoint source pollution.

EPA defines resource extraction as "mining, petroleum drilling, and runoff from mine tailing sites," and reports that resource extraction is responsible for the impairment of 11% of the nation's rivers. Mining activities cause water pollution in the form of sediment, acid drainage, metals, heat, channel alteration, and habitat modification.

While livestock grazing, timber production, and resource extraction are three major causes of water pollution originating on public lands, there are many less widespread nonpoint source activities that cause water pollution. Other pollutants reported by EPA include salts, oil, and grease. Salt contamination can impair freshwater ecosystems, making them unfit for both aquatic life and some human uses. For example, in Mount Hood National Forest, the Timberline ski area has a permit to use salt to keep its summer ski trails firm. In the summer of 1995, 1.2 million pounds of salt were applied to a 64 acre snowfield which is the headwaters for much of Mount Hood's drainage. This is just one example of a federally permitted nonpoint source activity that is particular to a certain land use.

Regulation of nonpoint source activities on public lands would bring us a step closer to achieving water quality standards across the nation. Professor William Rodgers writes:

In the forefront of any analysis of nonpoint source pollution is the role of the federal government as manager of approximately one-third of the U.S. land area constituting the federal lands. The federal government is intimately involved in a wide-range of activities that include the development of energy sources on federal lands (oil shale, coal, uranium and geothermal sources) as well as a number of other practices (grazing, silvicultural, construction of roads, railroads, airports, siting of landfills) that can lead to water pollution problems.

It seems that the expectation for our public lands and waters to accommodate high-impact industries such as livestock grazing, silviculture, and mining cannot be reconciled with the CWA goals of fishable and swimmable waters and "biological integrity." If states plan to take seriously the CWA mandate to protect the existing uses of their waters from nonpoint source pollution, they must address the federal land use practices that jeopardize the fish, wildlife, and human needs that depend upon healthy aquatic ecosystems.

The Scope and Strength of Section 401 of the CWA

The scope and strength of section 401 of the CWA presents an opportunity for states to regulate the broad range of federal activities that cause nonpoint source pollution, and require the federal government to comply with state water quality standards.

How Does Section 401 State Water Quality Certification Work? Section 401 of the CWA gives states the authority to grant or deny "certification" for "any federally permitted activity which may result in a discharge into the navigable waters."

In deciding whether to grant or deny certification, the state considers whether the proposed activity will comply with state water quality standards (and several other CWA provisions). For an activity to be subject to the required certification of section 401, it must meet the following two threshold conditions: the activity must require a federal permit, and it must potentially result in a discharge.

By granting water quality certification, a state is saying that the proposed activity will not violate state water quality standards. If a state denies certification because the proposed activity will not comply with water quality requirements, the federal permitting agency is prohibited from issuing the permit or license. Section 401(d) gives states the option of placing conditions on the certification to assure compliance with water quality standards and these conditions are then incorporated into the federal permit. By failing to act on a certification request within one year, the state forfeits its authority and the certification requirement is waived.

The Scope of Section 401 State Certification

The concept of state water quality certification first appeared in 1970, but state implementation of water quality certification authority has been a slow evolutionary process. Initially, state certification was applied exclusively to only a few federal permits, Major confusion arose when state and federal court decisions conflicted as to the nature and extent of the conditions that a state could include in its certification. In 1994, however, the Supreme Court handed down PUD No.1 of Jefferson County v. Washington Department of Ecology. This landmark case resoundingly confirmed the idea that state certification can include conditions limiting not just the "discharge," but any aspect of the activity that might affect water quality.

State Application of Section 401

Historically, states exercised their certification authority almost exclusively in response to federal hydroelectric projects, point sources requiring National Pollutant Discharge Elimination System (NPDES) permits, and discharges of dredge and fill material under section 404 of the CWA. However, the statute does not expressly state that section 401 applies to only these federal permits. On the contrary, the statute says that "any applicant for a Federal license or permit to conduct any activity" which may cause a discharge must receive state certification. However, the statute does not specifically define these terms, nor does it provide a list of specific permits. After analysing the discrepancy between the broad statutory language and the states' narrow application of it, several commentators have concluded that section 401 could potentially apply to a broad range of federal permits and licenses.

For example, state certification might be required for the issuance of timber contracts, deepwater port permits, Endangered Species Act incidental-take permits, and Clean Air Act permits. In The Untapped Power of Clean Water Act [sections] 401, Debra Donahue argues that "the universe of 401-reviewable permits" is bounded only by "the statutory prerequisite of a `discharge into the navigable waters." No court has yet overturned a state's decision to require certification of an activity on the grounds that the "Federal license or permit" applied for was outside of the state's section 401 jurisdiction, and recent judicial opinions have not focused on the license or permit language of the Act.

Certification Conditions

Rather than denying certification of an activity, states have typically exercised their right to include conditions in the certification which are meant to ensure that the activity will not violate water quality standards or effluent limitations. Section 401(d) gives states the authority to place any conditions on a water quality certification that are necessary to guarantee compliance with water quality standards, effluent limitations, standards of performance, pretreatment standards, more stringent state law requirements, or "any other appropriate requirement of State law." For example, in 1982, Maine imposed oil-spill prevention and water-quality standard requirements on a proposed oil refinery. Similarly, in 1992, Vermont conditioned certification of a hydroelectric project on the permittee's protection of aesthetic and recreational values of the water body. In 1986, an Oregon state court upheld Oregon's use of land use restrictions as certification conditions to the extent that the state could show a connection between the conditions and a project's compliance with water quality standards. Nevertheless, conditions imposed on state certifications were often the subject of appeal. Further, courts were undecided as to whether entire projects could be subject to conditions and as to the type of conditions which were authorised by section 401.

In 1994, the Supreme Court answered the questions as to the proper scope of certification conditions in PUD No. 1. Prior to 1994, the State of Washington imposed minimum stream flow requirements on its certification of a hydroelectric project. In response, the local utility board argued that section 401 conditions could be based only on specific chemical or numeric criteria, and that any certification conditions had to be directly associated with a point source discharge of pollutants. Relying to a great extent on the language of section 401(d), the Supreme Court held that a state can require compliance with all of its water quality standards, including designated uses, water quality criteria, and antidegradation regulation. Accordingly, the Court concluded that Washington's minimum stream flow requirement was a proper certification condition.

In addition, the Court held that once it is shown that an activity may result in a discharge, a state can impose conditions on the activity as a whole, and not just the "discharge." In PUD No. 1, both parties agreed that the dam would result in "two possible discharges," namely the release of dredged and fill material during construction, and the discharge of water at the end of the tailrace after its use to generate electricity. The instream flow conditions included in the state certification were not directly connected to either of the possible discharges. Section 401(d) provides that by certifying a federally permitted activity, the state is assuring that the activity will comply with state water quality standards, effluent limitations, and other requirements. Based on this language, the Court reasoned that a state could not be expected to confidently assure compliance with so many stringent requirements if it were not also given the authority to place conditions on all aspects of the activity.

The combined holdings of PUD No. 1 clarify the role that section 401 plays in the CWA regulatory scheme. State authority to require compliance with all portions of its water quality standards combined with state authority to place conditions on the whole of an activity make section 401 a powerful tool for states. However, PUD No. 1 did not directly address what has emerged as the fundamental unresolved question regarding the scope of section 401: do the requirements of section 401 state water quality certification apply to federally permitted nonpoint sources?

The Ninth Circuit Split

Two district courts in the Ninth Circuit recently addressed the question of whether section 401 applies to federally permitted nonpoint sources, and the courts reached directly conflicting conclusions. In Oregon Natural Desert Ass'n v. Thomas (ONDA), Judge Ancer Haggerty for the U.S. District Court of Oregon examined whether the term discharge in section 401 applied to the pollution resulting from cattle grazing under a permit issued by the United States Forest Service (USFS). The permit at issue in ONDA was first granted in July of 1993, and enabled defendant-intervenor Robert Burril to graze his cattle in the Malheur National Forest on an allotment that straddles the Middle Fork of the John Day River and Camp Creek (a tributary of the John Day River). The plaintiffs, a coalition of local environmental organisations, joined by the Confederated Tribes of the Warm Springs Reservation of Oregon as plaintiff-intervenors, filed suit against the USFS for erroneously issuing a grazing permit without first getting state certification that the permit would not violate water quality standards on the John Day River and Camp Creek. Plaintiffs requested a declaratory judgment that states certification under section 401 is required before a cattle grazing permit may be issued. They also sought an injunction prohibiting the USFS from issuing a federal cattle grazing permit unless and until the applicant obtains section 401 state certification.

On a motion for summary judgment, Judge Haggerty found in favour oft he environmental plaintiffs and the Tribes, granting both declaratory and injunctive relief. Based on the plain meaning of the word "discharge" as used in section 401, the statutory definition of "discharge" in the CWA, and the legislative history of the CWA, Judge Haggerty concluded "that [sections] 401 applies to all federally permitted activities that may result in a discharge, including discharges from nonpoint sources." The opinion notes that a plaintiff need only show that the offending discharge might result in pollution, and the plaintiffs in ONDA offered undisputed proof that cattle grazing on the Camp Creek allotment resulted in and would continue to cause pollution of both Camp Creek and the Middle Fork of the John Day River. The court held that the pollution caused by cattle grazing on the Camp Creek allotment was a "discharge... into navigable waters' within the meaning of [sections] 401 of the CWA" and therefore certification of the permit by the State of Oregon was required. Judge Haggerty did not focus on a specific discharge caused by the grazing to find that it required section 401 certification; instead, he focused on the undisputed evidence before the court that cattle grazing on the Camp Creek allotment caused water pollution in both Camp Creek and the John Day River.

In Idaho, Judge Mikel H. Williams for the U.S. District Court of Idaho considered whether the term discharge in section 401 includes pollution caused by a logging road that was approved by the United States Forest Service. Plaintiff, the Idaho Conservation League (ICL), challenged the Goat Roost Roads Proposal Decision (Roads Proposal) which had been signed by the supervisor of Clearwater National Forest, defendant James Caswell. The Roads Proposal gave defendant-intervenor Plum Creek Timber Company permission to build two roads in Clearwater National Forest for the purpose of providing Plum Creek with access for logging on its private property. The roads approved in the Roads Proposal run parallel to, and on either side of, Walton Creek, a water body which both parties concede provides habitat for steelhead trout, bullhead trout, cutthroat trout, salmonid spawning, and a spring chinook salmon hatchery. ICL alleged that the Forest Service's failure to obtain state certification of the Roads Proposal violated section 401 of the CWA, because construction of the roads would cause discharge of sediment and other pollutants into Walton Creek. ICL offered evidence that past logging activities and road construction on Walton Creek has caused an accumulation of sediment in violation of water quality standards, and that construction of the two new roads would likely result in further erosion and excess sediment, which in turn would degrade fish habitat.

On a motion for summary judgment, Judge Williams concluded that section 401 certification was not required for the Roads Proposal. The court did not need to determine whether construction of the roads might result in the alleged discharge, because it found that the statutory definition and context of section 401 in the CWA made it clear that it was intended only to apply to activities which result in a point source discharge, and a logging road is not a point source. Both Ninth Circuit district courts considered whether the term discharge as it is used in section 401 includes discharges from nonpoint sources, but the courts came to directly conflicting conclusions. The ONDA court interpreted the term discharge to include discharges from nonpoint sources, while the Caswell court's interpretation limited section 401 to discharges from point sources.

By looking at the plain meaning of discharge, the context of section 401 in the CWA regulatory scheme, legislative history, and federal water pollution control policies, the next section of the Comment attempts to discern the proper interpretation of discharge in section 401. For the sake of clarity, this section outlines the strongest arguments both for and against a broad interpretation of "discharge." An objective discussion of the plain meaning, statutory context, and legislative history of the term discharge as it is used in section 401 follows in the conclusion of this section.

Arguments in Support of a Broad Reading of the Term "Discharge"

The Definition of "Discharge" in the CWA

The plain meaning of discharge in section 401 is necessarily guided by the parameters of the CWA's statutory definitions. The 1972 House Report said of the CWA, "in order to fully understand the Act, it is necessary to recognise that certain terms... have very specific and technical meanings." First, it is helpful to remember that the full phrase at issue in section 401 is that state certification is required for any federally permitted activity which "may result in any discharge into the navigable waters." The CWA provides that "the term discharge when used without qualification includes a discharge of a pollutant, and a discharge of pollutants." The Act further provides thai: "the term discharge of a pollutant and the term discharge of pollutants each means... any addition of any pollutant... from any point source." This definition is repeated in EPA regulations, and the Act offers no further definition. The CWA definition can be read as stating that discharge, when used without qualification, includes discharge from point sources. Because the only word preceding the term discharge in section 401 is "any," section 401 uses the term discharge without qualification.

Much of the reasoning of the ONDA opinion relies on the use of the term "includes" in the CWA definition of discharge. The word "includes" is a "term of enlargement, not of limitation, and... the reference to certain entities or categories is not intended to exclude all others." The use of the term "includes" in a statutory definition indicates that the definition that follows is a non-exclusive list that can be enlarged, and that items not clearly listed in the definition may nonetheless be covered by the term. The significance of the congressional choice to use the term "includes" rather than "means" in the definition of discharge is highlighted by the fact that of the twenty terms defined in section 502 of the CWA, nineteen of the definitions employ the word "means," and only the definition of discharge uses "includes." The argument goes that unless we presume that Congress's use of the term "includes" was the result of careless drafting, it seems that Congress intentionally left the definition of discharge open to nonpoint sources of discharge. One common sense reading of the definition of discharge in the CWA is that a discharge includes, but is not limited to, discharges from point sources. Arguably, to give "includes" the same meaning as "means" not only confuses the English language, but also makes a mockery of careful legislative drafting.

The CWA definition of discharge makes it clear that the term, when used without qualification, is not limited to discharges from point sources. However, the definition of discharge in the Act does not affirmatively state what kinds of water pollution the term does apply to. To fill in this gap, it is proper to turn to a canon statutory construction: unless Congress gives a statutory definition, the ordinary usage of a term should control. In the CWA, Congress provides a definition of the term discharge, but it gives only a partial definition. Therefore, it is appropriate to use the ordinary usage of the term discharge to complete the definition of discharge as it is used in section 401 of the CWA. A discharge is defined by Webster's Dictionary as "a flowing or issuing out; something that is emitted."

Arguably, the CWA definition of discharge indicates that Congress recognised this ordinary meaning of discharge, yet wanted to be sure that when the Act refers to discharge (without qualifying it as discharge of a pollutant), courts would interpret the term to include discharge from point sources. The CWA definition of discharge is phrased to clarify the inclusion of point source pollution in the term discharge, not to exclude any ordinary meaning that it has. The statutory definition of the term discharge in the CWA combined with the ordinary usage of the term provides support for the argument that section 401 of the CWA applies to nonpoint sources as well as point sources.

The Regulatory Scheme of the CWA

The U.S. Supreme Court has stated that each provision of a statute should be read in reference to the whole act. The CWA is a complex statute made up of hundreds of provisions, each of which is essential to achieving Congress's water pollution control goals. At the heart of the CWA regulatory scheme is the concept of shared administration of water pollution control programs between the federal government and the states. Section 101 declares that "it is the policy of Congress to recognise, preserve, and protect the primary responsibilities and fights of States to prevent, reduce, and eliminate pollution and to plan the development and use... of land and water resources." Recognition of the primary rights of states to address water pollution within their borders is central to the CWA. A careful look at several major provisions of the Act suggests that Congress intended for states to govern their own water quality, limited only by EPA's setting of water quality standard and program minimums. Application of section 401 to nonpoint sources gives states the opportunity to insure that federally permitted activities not subject to the NPDES program will be in compliance with state water quality standards. This application of section 401 is consistent with the statutory scheme of the CWA, because the states should have the right to "prevent, reduce, and eliminate" water pollution caused by federally permitted activities.

Requiring federally permitted nonpoint source activities to be certified by the state for compliance with state water quality standards is in harmony with section 313(a) of the Act, which directs the federal government to comply with state water quality standards. Section 313 requires all agencies "engaged in any activity resulting, or which may result, in the discharge of runoff of pollutants . . . shall be subject to... all... State... requirements.., respecting the control and abatement of water pollution ...." Significantly, the Ninth Circuit has consistently interpreted this statutory provision as requiring the federal government to ensure that nonpoint source activities on federal land comply with state water quality standards. Sections 313 and 401 were originally drafted together to give states the authority to require both federal activities and federally permitted activities to comply with state water quality standards. Application of section 401 to nonpoint sources gives states nothing more than a check on the federal government's obligation under section 313 to comply with water quality standards.

Under the CWA statutory scheme, states are to protect existing and beneficial uses by enforcing their water quality standards. For example, section 303 of the Act requires states to "identify stream segments which do not meet applicable water quality standards." The state must prioritise these "water quality limited segments" (WQLS), and calculate a "total maximum daily load" for WQLS designed to bring the water quality back into compliance with the relevant standard. The section 303(d) list requirement of the CWA is another illustration of the role that states are to play in regulating nonpoint source pollution for compliance with state water quality standards. Without the authority to certify federally permitted nonpoint source activities to comply with state water quality standards, states would be missing an essential tool that the statutory scheme of the CWA suggests they should have to achieve water quality standards.

Yet another aspect of the CWA regulatory scheme suggests that section 401 does not apply only to point sources. Under the CWA, all point sources are required to get either a NPDES permit or a section 404 dredge and fill permit. If section 401 applies only to point sources, it would seem more appropriate to include the state certification requirement as a subsection to the point source provisions of the CWA. Instead, section 401 occupies its own unique niche in the CWA regulatory universe, and has authority over "Any applicant for a Federal license or permit to conduct any activity ...."

The Legislative History of the CWA

Section 401(a) originated as section 21(b) of the Water Quality Improvement Act of 1970, and the language of section 401 remained identical to section 21(b) through the 1972 enactment and all subsequent amendments of the CWA. It is important to remember that when section 21(b)) was first enacted, the federal water pollution control program regulated all polluting activities through water quality standards. In the original Federal Water Pollution Control Act, there was no distinction between point and nonpoint sources, and no definition of the term discharge as it was used in section 21.

Section 21(b) of the Water Quality Improvement Act of 1970

At the time of passage of section 21(b), Congress was fully aware of the water quality impacts from nonpoint source pollution. The House Report stated that section 21(b) would require certification of "any activity of any kind or nature which may result in discharges into the navigable waters." This broad language suggests that the modern section 401 does apply to nonpoint sources, because "any activity of any kind" is language clearly inclusive of activities like grazing, clearcutting, and mining that cause nonpoint discharges. In fact, "any activity of any kind" leaves very little room for exceptions. Referring to the wide variety of permits and licenses issued by the federal government, the House Report on section 21(b) further stated that "many federally issued permits involve activities or operations potentially affecting water quality. The purpose of this subsection, is to provide reasonable assurance... that no license or permit will be issued by a Federal agency for an activity that... could in fact become a source of pollution." The language of the report suggests that the drafters of section 21(b) were focused on giving states authority to regulate any kind of federally permitted activities that might cause water pollution.

The "any activity" language is present in the Senate Report and throughout the legislative history of section 21(b). The Senate Report reflects that the purpose of the bill was to ensure that the federal government would practice what it preached, by requiring "all Federal activities to comply with the philosophy and intent of the Nation's water quality program." The Senate Report notes that "in the past, these licenses and permits have been granted without any assurance that the standards will be met or even considered." Throughout the congressional hearings, commentary was focused on the general goal of the whole of section 21 to conform all federal activity with water quality standards, through the combined efforts of sections 21(a) and 21(b). Section 21(b) applied to all activities under direct federal control, while section 21(a) applied to activities licensed or permitted by the federal government. Considering that the general legislative goal of section 21 of the Water Quality Improvement Act (WQIA) was to give states the authority to regulate discharges from activities involving the federal government, it seems that section 21(b) was intended to apply to all federally permitted activities that might affect water quality, regardless of whether the source would now be considered a point or nonpoint discharge.

The Conference Report states that "applicants... seeking a Federal license or permit to conduct any activity of any kind or nature which may result in discharges... must obtain certification." Many members of Congress expressed an understanding that the focus of section 21 was to insure that federal activities, including federally permitted activities, do not contribute to water pollution. Senator John S. Cooper (R-Ky.) stated that the 1970 amendments "require, without exception, that all Federal activities that have any effect on water quality be conducted so that water quality standards be maintained." Representative Edmond A. Edmondson (D-Okla.) spoke regarding section 21(b), and observed that "a wide variety of licenses and permits... are issued by various Federal agencies... and many... involve activities... potentially affecting water quality," and the purpose of section 21(b) is to remedy that situation, In fact, the legislative history of the WQIA of 1970 suggests that at least some members of the legislature intended for section 21(b) to regulate federal activities resulting in water pollution, regardless of whether or not there is even a discharge. The 1970 legislative history of section 21 does not provide evidence that its application was intended to be limited to discharges from discrete conveyances, later known as point sources. On the contrary, it strongly suggests that Congress intended for section 21 to require state certification of any and all federally permitted activities potentially causing water pollution.

The 1972 Amendments to the CWA

Even though the passage of the 1972 Amendments centred around regulation of point source pollution, the legislative history does not reveal any congressional intent to limit state certification of federally permitted activities to point sources. An argument can be made that Congress meant for section 401 to shift its focus to point sources along with the rest of the Act. However, the Ninth Circuit Court of Appeals has determined that the introduction of point source regulation by effluent limitations was intended as an addition to the water quality standards program, not a replacement, In Northwest Environmental Advocates (NWEA) v. City of Portland, the court observed that "nowhere does Congress evidence an intent to preclude the enforcement of water quality standards that have not been translated into effluent discharge limitations .... [Congress's 1972 Amendments to the CWA were] intended to improve enforcement of pollution from point sources, not to supplant the old system." Similarly, a letter from the then Environmental Protection Agency administrator regarding the 1972 amendments stated, "we must build a system that includes the wise concepts of the past and build upon them .... The new law must build upon the existing foundation of water quality standards and employ effluent limitations as a tool for the achievement of those standards."

The technology based effluent limitations and water quality standards are equally important prongs of the CWA regulatory scheme. The argument that the entire focus of the Act shifted to point sources in 1972, and took section 401 with it, is weakened by the recent case law interpretations reflecting the importance and coexistence of both prongs of the CWA. Rather, it seems that the purpose of section 21(b) was to regulate all water pollution which might result from federal activities to meet water quality standards, and this purpose was not altered by the 1972 addition of point sources and effluent limitations to the regulatory scheme. As to the impact of the 1972 Amendments on section 401 more particularly, legislative history arguably confirms that Congress intended for the original purpose and application of section 401 to be preserved. For example, in the Senate debate Senator Muskie (D-Me.) stated, "all we ask is that activities that threaten to pollute the environment be subjected to the examination of the environmental improvement agency of the State for an evaluation." The thrust of legislative interest in section 401 again seemed focused on regulation of federal activities that might result in pollution, not on distinctions between point and nonpoint sources.

Federal Water Pollution Control Policy

Courts generally turn to policy considerations to supplement their understanding of a statute, For the purposes of this discussion, it is also helpful to take a step back to look at the underlying purpose of the CWA and federal water pollution policy. If states can use their section 401 authority to authorise and place conditions on federally permitted activities like livestock grazing, logging, and mining, most states will find themselves significantly closer to fulfilling the mandates of federal water pollution control policy. By giving states the authority to regulate federally permitted land use activities, section 401 could help states to successfully incorporate important principles of watershed management into their water pollution control programs.

Following Federal Water Pollution Control Policy

Application of section 401 to all federally permitted activities that may cause nonpoint source pollution is arguably a natural and necessary step towards achieving the goals of the CWA. The overriding purpose of the CWA is to "restore and maintain the chemical, physical, and biological integrity of the Nation's waters," with an eye towards protection of designated uses such as fish and wildlife habitat and recreation, Research demonstrating the seriousness of nonpoint sources of water pollution coupled with an increased understanding of ecological principles has placed increased regulation of nonpoint sources high up on the national water pollution control priority list. State regulation of federally permitted nonpoint source pollution complements the authority given to states by sections 208 and 319 of the CWA.

Federal water pollution control policy focuses on attainment of state water quality standards; applying section 401 to nonpoint sources would give states an effective new tool for achieving that goal. If section 401 applies to all discharges causing pollution, states have the authority to take some markedly new directions in their management of activities that cause nonpoint source pollution. PUD No. 1 gives states the power to place any condition on certification of a permit that will assure that the activity will not cause a violation of water quality standards. In the nonpoint context, the PUD No. 1 holding means that a state can go as far as it wants to protect water quality from nonpoint source pollution. If a state is serious about protecting or restoring water quality of a given water body that would be affected by a proposed activity, section 401 would give the state the authority to deny certification of the activity, and thereby bar issuance of the permit. Alternatively, states could simply adopt the best management practices (BMP) that are already recommended for certain land use activities. For example, states could condition certification of a timber sale on implementation of existing BMPs, such as establishment of streamside buffer zones, use of erosion prevention, and control measures and revegetation in the harvest area. Federal land management agencies are already required to include BMPs in permits to satisfy section 313 of the CWA. One advantage of state certification conditions would be increased flexibility, as they could vary from permit to permit, from month to month, and from watershed to watershed. Existing BMP requirements are relatively static and drafted to apply in a very generalised and discretionary way. State certification conditions, on the other hand, could potentially reflect scientific advances in management practices, local public opinion on resource and water management, and state concerns about particular water bodies.

Watershed Management—an Ecological Rationale

Just as our understanding of water pollution has evolved, so has our understanding of aquatic ecosystems and how to best preserve their biological integrity. Most current research indicates that if we want to protect and restore water quality, we must maintain the health of the entire aquatic ecosystem, Similarly, the aquatic ecosystem of a certain stream segment is only as healthy as the watershed to which it is connected. As pollution control practices integrate and adopt principles of watershed management, it will be important for one centralised entity to be responsible for tracking and regulating the multitude of land use activities taking place in each watershed. States are a natural nerve centre for monitoring the health of the waters within their borders. In addition to regulating and tracking the private and state activities having impacts on each watershed, states should be authorised to regulate any federally permitted activities that may result in discharges.

Because nonpoint source pollution is the result of land use and development decisions, it is imperative that regulation of nonpoint sources be approached by states and based on watershed management principles. Adoption of watershed management principles into water pollution control programs could be facilitated by application of section 401 to nonpoint sources of pollution because section

401 compels an intersection of land and water management. Successful watershed management will require that land managers and water quality managers work together to protect and maintain ecosystem integrity. By authorising states to condition federally permitted activities on compliance with water quality standards, section 401 could be an instrument for implementing watershed management goals.

Arguments in Favour of a Narrow Reading of the Term "Discharge"

Use of the Terms "Discharge" and "Runoff" in the CWA

The definition of discharge in the Act makes it clear that the term is not limited to discharges of pollutants from a point source. However, one can argue that using the word "discharge" instead of the phrase "discharge of pollutants" indicates that the provision applies to discharge from a point source, but not of material listed as a pollutant in the CWA. Under this argument, the fact that Congress did not limit the term to point source discharges of pollutants does not mean that it intended for the term to be applied to diffuse surface runoff and other forms of nonpoint source pollution.

Examination of Congress's use of the terms discharge and runoff in the CWA suggests that Congress generally did not use discharge to refer to nonpoint source pollution. If Congress intended for section 401 to apply to diffuse nonpoint source pollution, it could have used the term runoff instead of discharge. In fact, Congress did use both words in a section of the CWA governing federal facilities: "Each federal facility engaged in any activity resulting ... in the discharge or runoff of pollutants ... shall be subject to all water pollution laws." This provision was first drafted in 1970 as section 21(a) of the Water Quality Improvement Act of 1970, and it originally referred to "any discharges allegedly contributing to pollution from any such Federal property ... "in its mandate to conform all federal activities with water quality standards. In 1972, section 21(a) was transformed into section 313 of the CWA (Federal Facilities Pollution Control), and Congress amended "any discharge" to read discharge or runoff of pollutants. This careful drafting by Congress to insure that section 313 applies to both point and nonpoint sources suggests that it could have amended section 401 similarly had it wanted to.

Another argument is that the term discharge, as it is used in section 401, must refer only to discharges from point sources because roughly nine out of every ten times that it appears in the CWA, discharge is used to refer to point source discharges. Since Congress apparently uses the term runoff to describe pollution from nonpoint sources, nothing in the CWA suggests that discharge should be read broadly to include nonpoint sources of pollution. For example, section 208 refers to agricultural and mining pollution as runoff, and the word discharge is never used. In section 319, nonpoint source pollution is referred to as pollutant loadings, and section 304 of the Act refers to nonpoint source pollution as runoff. Only in section 105 of the Act did Congress include runoff in the meaning of discharge. Section 105 requires the Administrator to "develop, refine, and achieve practical application of... waste management methods applicable to point and nonpoint sources of pollutants to eliminate the discharge of pollutants, including, but not limited to, elimination of runoff of pollutants and the effects of pollutants from accumulated sources." In this provision, Congress seems to have included runoff in the term "discharge of pollutants," a term expressly defined to mean discharge from a point source. Use of the terms runoff and discharge in the Act suggests either 1) that Congress has generally used runoff when referring to nonpoint source pollution, and therefore discharge should not be read broadly to include runoff, or 2) Congress's use of the terms has not been consistent or logical enough to imply an intent that discharge not include runoff.

Congressional Acquiescence

Although Congress has amended the CWA several times since 1970, focusing on nonpoint source pollution in 1987, it has never amended section 401 to correct or clarify how it was being applied by states, courts and federal agencies. Until the ONDA case, section 401 had only been applied to hydroelectric projects, section 404 dredge and fill permits, and point source NPDES permits. If section 401 was actually intended to apply to federal permits for both point and nonpoint source activities, it seems likely that Congress would have amended the provision to insure proper implementation.

The CWA Statutory Scheme and Practical Concerns

Another argument against a broad reading of the term discharge is that since Congress has drafted specific provisions of the CWA for the exclusive purpose of regulating nonpoint source pollution, application of section 401 to the nonpoint source pollution problem would create a duplication of regulatory control. While the existence of CWA provisions to regulate nonpoint source pollution is not very convincing evidence for the proposition that Congress did not intend for section 401 to apply to nonpoint source discharges, it does raise some difficult implementation issues. If section 401 is applied to any nonpoint source activities which may result in a discharge, it is possible that the administrative burden on state agencies, private parties, and federal agencies might outweigh actual improvements in water quality. Would section 401 certification of federally permitted nonpoint source activities achieve any water pollution control that cannot be achieved through another CWA provision or existing means? Arguably not. Under section 319 of the CWA, states are already required to apply BMPs to nonpoint source activities affecting water quality. Many federally permitted activities are already required to comply with water quality standards and BMPs, so that application of section 401 to nonpoint sources will only require additional time, money, and paperwork for all parties.

While nonpoint source pollution from federally permitted activities is a serious problem that needs to be regulated for compliance with state water quality programs, there are arguably more efficient ways to approach the problem than state certification. For example, a key component of EPA's most recent draft strategy on nonpoint source management is to "strengthen federal partnerships." "To promote federal consistency with and support for State programs," EPA proposes Memorandums of Understanding (MOUs) between both EPA and federal agencies, and between states and federal agencies. Like section 401, these MOUs could be used to bring federally permitted nonpoint source activities into compliance with state water quality standards. Where section 401 would require state certification for each and every federal permit or license, the MOUs might offer a more streamlined and efficient approach to achieving the same goal: federal compliance with state water quality standards.

Furthermore, if the term discharge as it is used in section 401 is interpreted to mean pollution more diffuse than a point source yet not inclusive of all nonpoint source pollution, application of section 401 to all federally permitted activities might have absurd and tedious results. If discharge means a flowing or issuing out, we are signing courts up for hour upon hour of line-drawing because the definition is extremely vague. Would the line ever be drawn the same way twice? How would a court deal with typical nonpoint source activities, such as road-building, which may cause pollution but not a discharge? Application of section 401 to nonpoint source discharges would require identification of the sub-category of nonpoint sources that can be considered "discharges" subject to section 401 certification. Would section 401 state certification be required before the Forest Service could issue a backcountry backpacking permit? Professor Rodgers has commented, "for simplicity, the universe of the causes of water pollution should be considered as covered fully by the categories of point and nonpoint sources." In the case of section 401 discharges, simplicity is a strong argument against a broad reading of the term "discharge." Arguably, application of section 401 to nonpoint sources will have such absurd results that courts should not give it this construction.

Application of Section 401 to Nonpoint Sources is Generally Supported

The CWA is ambiguous and contradictory in many respects, but on balance, the plain meaning, legislative history, and federal policies behind section 401 generally support a reading of the term discharge that includes discharges from nonpoint sources. An objective look at the plain meaning and statutory definition of the term discharge reveals arguments both for and against a broad reading of discharge that includes nonpoint source discharges. On balance, the logic of the arguments for a broad reading are more persuasive and true to the canons of statutory interpretation. It is true that limitation of the term discharge to point sources is supported by the fact that nine out of every ten times Congress uses the term discharge in the CWA, it is referring to a point source discharge. The statutory definition of the term discharge also focuses on point source discharges. However, the fact that the definition of "discharge when used without qualification" is the only one of twenty definitions in the CWA that uses "includes" instead of "means" is strong support for a broad reading of the term.

Furthermore, when Congress first drafted section 21 in 1970, water pollution control terminology had not yet evolved to distinguish between point and nonpoint sources, much less between discharge and runoff. The phrase "any discharge" as it is used in section 401 should be interpreted to mean any discharge, whether from a point or nonpoint source. If a federally permitted nonpoint source activity may result in a discharge, plain meaning and the statutory definition support requiring section 401 certification that the activity will not violate state water quality standards. Application of section 401 to nonpoint sources is also generally supported by the CWA statutory scheme, which recognises that states have the primary rights and responsibilities to reduce and prevent water pollution within their borders.

The Caswell court reasoned that application of section 401 to the nonpoint source pollution problem would create a duplication of regulatory control since Congress has drafted specific sections to exclusively govern nonpoint source pollution. Sections 208 and 319 of the CWA give the states clear mandates and direction for regulation of nonpoint source pollution, and are the primary means through which the CWA regulates nonpoint sources of pollution. While the intent of the Caswell court was to interpret section 401 in a way consistent with the structure of the CWA, it seems to misunderstand the true nature of both the CWA and section 401. There can be no doubt that a logging road is considered a nonpoint source of pollution by the CWA, and is therefore subject to the nonpoint source provisions of the Act. However, nothing in the language or the structure of the CWA suggests that sections 208 and 319 are the only provisions through which nonpoint sources can be regulated. Nor is there any suggestion that section 401 can be applied only to point sources. The purpose of the Act is to "restore and maintain" the nation's waters, and the Act incorporates many provisions and approaches to achieve this goal.

There are provisions specifically regulating oil, sewage, toxics, and even pollution from federal facilities and properties, Section 401 regulates pollution from federally permitted activities. Section 401 does not derive its jurisdictional authority from whether a source is point or nonpoint, but from whether there is a federally permitted activity that may result in a discharge. While it is true that the CWA regulates point and nonpoint sources distinctly in some provisions, the regulatory scheme of the Act does not preclude one provision from regulating both point and nonpoint sources in the same manner. Where a statute uses multiple regulatory provisions to address the same problem, the statute should be read to give effect to all of its provisions. In fact, the clearest signal from Congress about regulation of nonpoint source pollution is that it should be left in the hands of the states. Application of section 401 to nonpoint source pollution is in harmony with that principle.

By giving states the opportunity to certify federally permitted nonpoint source activities for compliance with state water quality standards, section 401 can work in harmony with the other provisions of the Act requiring states to regulate nonpoint source pollution. The trade-off is that state certification of nonpoint source activities will place a new administrative burden on state and federal agencies as well as private permit applicants, who would argue that the state certification process is not the most efficient way for states to bring federally permitted activities into compliance with water quality standards. But the section 401 certification process is ultimately very flexible, and can be exercised by states as aggressively or passively as they choose. Interpretation of the term discharge in section 401 as applying to nonpoint sources is in harmony with the CWA statutory theme of state regulation of nonpoint source pollution, including nonpoint source pollution from federal activities.

The legislative history of the 1970 Act suggests that the intent of the legislature was for section 21(b) to require state certification for any federally permitted activity affecting water quality, without distinction between point and nonpoint sources. The argument that the 1972 amendments to the CWA shifted the focus of the entire Act, including section 401, has some merit. Both legislative history and recent caselaw highlight the importance of the water quality standards prong of the CWA. Would limiting section 401 to point sources really threaten the water quality prong of the CWA? Arguably not, because other provisions of the Act exclusively address regulation of nonpoint sources for compliance with state water quality standards. However, although the thrust of the 1972 amendments to the CWA was regulation of point sources, the legislative history of the 1972 Amendments to the Act does not reflect any clear legislative intent to limit application of section 401 to point sources.

Tools of statutory construction exist to assist a court in determining the intent of the legislature. The current Supreme Court tends to favour a textualist approach to statutory interpretation, which looks to "dictionary definitions, rules of grammar, and canons of construction, in an effort to derive the putatively objective meaning of the statutory word or phrase." Because neither the statutory definition nor the ordinary usage of the term discharge suggest that it applies only to point sources, it seems fairly clear that a textualist approach to determining the intended meaning of discharge would conclude that discharge includes nonpoint source discharges. Another jurisprudential approach to statutory interpretation is "intentionalism," which "refers to the use of a variety of tools, including legislative purpose and legislative history, in an effort to determine the intent of the legislature when it included a particular word or phrase in a statute." Under the intentionalist approach, the term discharge in section 401 should be applied to nonpoint source discharges because Congress intended for section 401 to apply to both point and nonpoint sources. Even in light of all the internal contradictions and ambiguities of the CWA regarding the proper definition of discharge, the holding of the ONDA court seems generally best supported. The plain meaning, statutory scheme of the CWA, and legislative history support Judge Haggerty's conclusion that the term discharge as it is used in section 401 includes discharges from nonpoint sources.

Section 401 Certification of Nonpoint Source Activities

Of course, it remains to be seen whether the Ninth Circuit will affirm the ONDA decision holding that section 401 of the CWA applies to both point and nonpoint sources. As Joe Hobson of the Oregon Farm Bureau said, "the Forest Service has one judge telling it to go south and the other north... This thing will go to the Supreme Court."

Yet even if the Supreme Court were to affirm the ONDA decision, there would be many new unanswered questions about state certification under section 401 of the CWA. Regulation of nonpoint source pollution is linked directly to land use management and many states may be uncomfortable taking an active role in modifying long-time land use activities to comply with water quality standards. Section 401 has the potential to partially move debates about timber and rangeland practices out of the national political realm and put the tough decisions into the hands of the states.

Inevitably, state decisions favouring water quality will disturb customary land use practices of local industries; states may find themselves making prickly decisions in pursuit of the elusive public interest. This tension may provoke some new questions. How hard do states have to look at the proposed activity? Can state certification mean issuance of a general permit and general conditions for a certain category of activity? Or must the state examine the water quality status of each affected water body and evaluate impacts caused by the specific proposed discharge?

The CWA does not give federal agencies the option of getting state 401 certification that federally permitted activities will not violate water quality standards; section 401 is a mandatory provision. The states, on the other hand, have the option of waiving the certification. To understand how application of the section 401 state certification provision to nonpoint source activities might play out in the real world, it is helpful to examine section 401 from the states' perspective and the Oregon Department of Environmental Quality's (DEQ) response to the ONDA decision.

Section 401 Certification of Nonpoint Sources from the States' Perspectives

From the states' perspectives, there are clear advantages to using section 401 to implement nonpoint source management. First, as soon as a state certification condition attaches to a federal permit, enforcement of that condition becomes the responsibility of the federal agency that issued the permit. While states might experience an increased workload and cost on the certification end of section 401, those costs would surely be outweighed by avoided enforcement expenses.

Because of the flexibility states will have in processing certifications and attaching conditions, it will be possible for states to create economically efficient procedures for certification reviews. Second, placing conditions on a permit for a discharge activity will allow the states to proactively prevent ecosystem degradation. Rather than helplessly watching the Forest Service issue a livestock grazing permit that impairs the quality of a certain stream segment which will later require costly state rehabilitation efforts, the state can simply attach conditions to the permit at the outset.

Common sense tells us that it is more cost-effective to regularly maintain a car engine than to pay for huge damages incurred by neglect; the same common sense reasoning applies to ecosystem management. Not only is it difficult and expensive to rehabilitate the aquatic ecosystems that promote water quality, but it also can take many years to undo damage caused by one clearcut or a season of grazing. From the states' perspective, application of section 401 to nonpoint source pollution offers some clear financial and ecological advantages.

Recognising the appeal of applying section 401 certifications to federally permitted nonpoint source activities, both Arizona and Washington have recently drafted rules to take advantage of the ONDA ruling. Arizona's "draft rule recognises grazing activities as nonpoint source pollutant contributors to surface water contamination," and gives specific requirements for certification by the state. Arizona plans to implement these rules unless and until ONDA is overturned.

The State of Washington has joined in with draft rules for regulating a broad range of nonpoint source activities under section 401, including road building and livestock grazing. Washington is gladly embracing the ONDA decision, stating that incorporation of watershed-based best management practices (BMPs) as conditions on section 401 certifications would require livestock grazing and road building permits to "guarantee that their activity will not contaminate waters." Arizona and Washington's draft rules indicate that some states would be eager to use state certification to regulate federally permitted nonpoint source pollution, but these rules remain in the draft stage.

A Look at Oregon's Response

Oregon Certification of Salting on Mount Hood—A Case Study

DEQ is the state agency in Oregon that handles section 401 certification reviews. After the ONDA opinion declared that federally permitted nonpoint source activities must be certified by the state of Oregon, DEQ was immediately faced with new situations. For example, Timberline ski area located on Mount Hood uses as much as 1.2 million pounds of salt each summer to keep the surface of the Palmer snowfield firm for best summer ski conditions. A local environmental group fried a lawsuit against Timberline, claiming that Timberline needed certification from DEQ to discharge salt into a public water supply.

The Forest Service had always considered salting a nonpoint source activity, and had never required Timberline to get DEQ certification of the release. While both the lawsuit and DEQ's decision on whether certification was required were pending, Timberline wisely decided that in light of the ONDA decision, it should apply for state certification. According to Deputy Director of DEQ Lydia Taylor, DEQ examined the impact that discharge of the salt was having on the affected watersheds, and determined that while data showed an increase in salt concentration over natural levels, water quality standards for salt were not close to being violated. DEQ certified salting of the Palmer snowfield, and the Forest Service issued the permit to Timberline. The salting on Mt. Hood scenario illustrates that state involvement in a federal land use permit can monitor water pollution without greatly disrupting existing practices.

DEQ "Universal 401 Certification" of Grazing Permits

On January 10, 1997, the Oregon Department of Agriculture (DOA) and DEQ responded to the ONDA opinion by proposing a "streamlined method of issuing section 401 certification to people who may need them in order to receive grazing permits from federal agencies." This shortterm response was the product of DEQ's concern that ranchers be able to get their grazing permits up and running by spring without going out of business. Since regulation of agricultural activities affecting water quality falls to the DOA under Oregon law, the DOA has drafted emergency rules that describe the conditions to be placed in a section 401 certification of a grazing permit. For its part, DEQ has issued draft emergency rules that amend its own procedural rules for

"Certification of Compliance with Water Quality Requirements and Standards" to include a new subsection called "Certification of Federal Grazing Leases." The two sets of rules are intended to work together to allow for a streamlined application process.

Oregon's DEQ response to the opportunity to issue state certification reviews of grazing permits has demonstrated remarkably little concern for assuring that each grazing permit they issue will not violate water quality standards. Instead, the main thrust of DEQ's response seems to be to maintain the status quo, and with as little paperwork as possible.

The DOA conditions that will be attached to federal grazing permits includes prohibition of "active channel erosion," and a requirement that "livestock grazing... within near-stream management areas shall maintain or improve appropriate vegetative cover to decrease or conserve stream temperature in the near-stream management area." The conditions incorporate discretionary, but sound ecological approaches to protecting water quality, and were adopted from the already existent DOA best management practices.

It is not the certification conditions proposed by Oregon that fall so far short of protecting water quality, but rather the state's "universal permit" approach to issuing the actual certification. DEQ's proposed rule provides that in lieu of individual permit applicants requesting state certification, "a federal land management agency may submit proposed conditions for certification of all its grazing leases within the state or a specified geographical area within the state.

If the conditions are accepted by DEQ and DOA, a general certification shall be issued for leases within the area covered by the agency's application." While this subsection certainly provides for an efficient and streamlined application process, it is not at all clear how effectively it will protect water quality standards. In January of 1996, DEQ published a list of 904 stream and river segments that do not meet state water quality standards, including streams and rivers from nearly all of the major watersheds in the state. Seven hundred thirty-eight water bodies have violated the temperature water quality standard. Other standards violated include bacteria and dissolved oxygen. Agriculture is considered a major cause of violations across the state. Considering that each water body is subject to different discharges and violations, it is difficult to understand how DEQ could issue generic certification conditions for grazing on BLM land across the state with any confidence that water quality standards will not be violated.

As National Wildlife Federation attorney Tom Lustig pointed out, "the states aren't there yet as far as a commitment to enforcing water quality standards.., they could easily say, `grazing has no effect, everything is OK.'" It remains to be seen how DEQ will handle section 401 certification reviews for other nonpoint source activities. Certainly if DEQ were pressured by public interest to attain water quality standards, section 401 would give DEQ the authority to strengthen certification conditions accordingly. It is clear, however, that it may take some time for states to realise the full potential of section 401 certification of federally permitted nonpoint sources affecting water quality. If affirmed on appeal, ONDA will represent a major shift in the regulation of nonpoint source pollution, and will have far-reaching implications. While it remains to be seen whether states will use section 401 certification as the powerful tool that it is for achieving water quality standards, states should at least be given the opportunity to certify nonpoint source activities under section 401. The plain meaning, statutory scheme, and legislative history of the CWA, as well as federal water pollution control policy, generally weigh in to support a broad interpretation of the term discharge. The Ninth Circuit should give full effect to the intent and goals of Congress by interpreting the term discharge as it is used in section 401 to include discharges from nonpoint sources.

3

Water Law, Water Politics and Water Crisis

1.6 billion people have gained access to a safe water source since 1990. The proportion of people in developing countries with access to safe water is calculated to have improved from 30% in 1970 to 71% in 1990, 79% in 2000 and 84% in 2004. This trend is projected to continue. To halve, by 2015, the proportion of people without sustainable access to safe drinking water is one of the Millennium Development Goals.

This goal is projected to be reached. A 2006 United Nations report stated that "there is enough water for everyone", but that access to it is hampered by mismanagement and corruption. In addition, global initiatives to improve the efficiency of aid delivery, such as the Paris Declaration on Aid Effectiveness, have not been taken up by water sector donors as effectively as they have in education and health, potentially leaving multiple donors working on overlapping projects and recipient governments without empowerment to act.

The authors of the 2007 Comprehensive Assessment of Water Management in Agriculture cited poor governance as one reason for some forms of water scarcity. Water governance is the set of formal and informal processes through which decisions related to water management are made. Good water governance is primarily about knowing what processes work best in a particular physical and socioeconomic context. Mistakes have sometimes been made by trying to apply 'blueprints' that work in the developed world to developing world locations and contexts. The Mekong river is one example; a review by the International Water Management Institute of policies in six countries that rely on the Mekong river for water found that thorough and transparent cost-benefit analyses and environmental impact assessments were rarely undertaken. They also discovered that Cambodia's draft water law was much more complex than it needed to be.

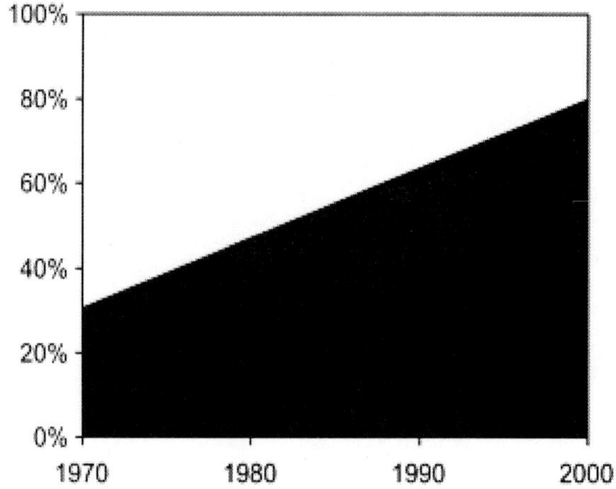

Figure: *Water politics is politics affected by water and water resources. For this reason, water is a strategic resource in the globe and an important element in many political conflicts. It causes health impacts and damage to biodiversity.*

The UN World Water Development Report (WWDR, 2003) from the World Water Assessment Program indicates that, in the next 20 years, the quantity of water available to everyone is predicted to decrease by 30%. 40% of the world's inhabitants currently have insufficient fresh water for minimal hygiene. More than 2.2 million people died in 2000 from waterborne diseases (related to the consumption of contaminated water) or drought. In 2004, the UK charity WaterAid reported that a child dies every 15 seconds from easily preventable water-related diseases; often this means lack of sewage disposal.

Organizations concerned with water protection include International Water Association (IWA), WaterAid, Water 1st, American Water Resources Association. The International Water Management Institute undertakes projects with the aim of using effective water management to reduce poverty. Water related conventions are United Nations Convention to Combat Desertification (UNCCD), International Convention for the Prevention of Pollution from Ships, United Nations Convention on the Law of the Sea and Ramsar Convention. World Day for Water takes place on 22 March and World Ocean Day on 8 June. Water used in the production of a good or service is virtual water.

Water Law

Water law is the field of law dealing with the ownership, control, and use of water as a resource. It is most closely related to property law, but has also become influenced by environmental law. Because water is vital to living things and to a variety of economic activities, laws attempting to govern it have far-reaching effects.

History

The history of man's relation to water illustrates varied approaches to the management of water resources. The Etruscans had a deep knowledge of hydrology and hydraulics, a knowledge which they put to good use in their many land drainage schemes. The lower lying portions of Rome such as the area between the Capitol and Velia was formerly marshland. Settlement of the low-lying ground would never have been a possibility without the hydraulic engineering skills of the Etruscans. This took place around 625 BCE when, according to archaeological evidence a network of drainage channels was dug through the marshy ground, and at the same time, the stream that separated the two hills of the Capitoline and Palatine was regulated, its embankments were strengthened, and it was finally covered over.

Without enclosure and drainage, more than half of the area of the present-day Netherlands would be flooded with every high tide, every wet season, or permanently. The struggle for the country's survival has largely determined its appearance and left numerous marks on it. In the triangle between the cities of Amsterdam, Haarlem and Leiden the 45,000-acre (180 km^2) Haarlemmermeer polder demonstrates an extraordinary step in the scale of land drainage. After centuries of preliminaries, a lake of unprecedented size was drained in three years (1849–1852) and transformed into valuable agricultural land. In 1899, construction of the first Aswan Dam was begun to address agricultural and energy shortages exacerbated by population growth in Egypt and the Sudan. Completed in 1902, its height was raised in subsequent building campaigns of 1907-12 and 1929-34. With the signing of the Nile Water Agreement by Egypt and the Sudan in November 1959, work began on the second Aswan dam. The second dam submerged much of Lower Nubia displacing 90,000 Egyptian peasants and destroying monuments and archaeological sites from the First to the Third Cataracts of the Nile River. The Aswan High Dam captures floodwater during rainy seasons and releases the water during times of drought. The dam also generates more than 10 billion kilowatt-hours every year. The project prevents the natural silting process which enriched Egyptian agriculture and farmers must now use about one million tons of artificial fertilizer as a substitute for natural nutrients that once fertilized the now arid floodplain.

Difficulties of Water Rights

Water has unique features that make it difficult to regulate using laws designed mainly for land. Water is mobile, its supply varies by year and season as well as location, and it can be used simultaneously by many users. As with property (land) law, water rights can be described as a "bundle of sticks" containing multiple, separable activities that can have varying levels of regulation. For instance, some uses of water divert it from its natural course but return most or all of it (eg. hydroelectric plants), while others consume much of what they take (ice, agriculture), and still others use water without diverting it at all (eg. boating).

Each type of activity has its own needs and can in theory be regulated separately. There are several types of conflict likely to arise: absolute shortages; shortages in a particular time or place; diversions of water that reduce the flow available to others; pollutants or other changes (such as temperature or turbidity) that render water unfit for others' use; and the need to maintain "in-stream flows" of water to protect the natural ecosystem.

One theory of history, put forward in the influential book *Oriental Despotism*, holds that many empires were organized around a central authority that controlled a population through monopolizing the water supply. Such a hydraulic empire creates the potential for despotism, and serves as a cautionary tale for designing water regulations.

Water law involves controversy in some parts of the world where a growing population faces increasing competition over a limited natural supply. Disputes over rivers, lakes and underground aquifers cross national borders. Although water law is still regulated mainly by individual countries, there are international sets of proposed rules such as the Helsinki Rules on the Uses of the Waters of International Rivers and the Hague Declaration on Water Security in the 21st Century.

Long-term issues in water law include the possible effects of global warming on rainfall patterns and evaporation; the availability and cost of desalination technology; the control of pollution, and the growth of aquaculture.

Water Law by Country

Canada: Under the Confederation Act, jurisdiction over waterways falls under the federal government. One federal statute is the "Navigable Waters Protection Act". Others are the "Arctic Waters Pollution Prevention Act" and the "Oceans Act".

At the provincial level, one piece of legislation is the Clean Water Act (S.O. 2006, Chapter 22) as enacted by the Provincial Parliament of Ontario Legislative Assembly, Canada. The purpose of this Act is to protect existing and future sources of drinking water within the province. Protecting water at its source is the first step in the[multi-barrier approach]] to source water protection. By stopping contaminants from getting into sources of drinking water — lakes, rivers and aquifers — the legislation provides the first line of defense in the protection drinking water.

Australia: Water law in Australia varies with each state.

Tasmania: A newly formed Tasmanian Water Corporation has compulsorily acquired all drinking water supply infrastructure without payment and does not have direct accountability.

Water Law in the United States

In the United States there are complex legal systems for allocating water rights that vary by region. These varying systems exist for both historical and geographic reasons. Water law encompasses a broad array of subjects or categories designed to provide a framework to resolve disputes and policy issues relating to water:

- Public waters, including watercourses have traditional included tidal waters and navigable waterway.
- Other surface waters—generally water that flows across non-public land from rain, floodwaters, and snowmelt before those waters reach public watercourses.
- Groundwater, sometimes called subterranean, percolating, or underground water
- Public regulation of waters, including flood control, environmental regulation—state and federal, public health regulation and regulation of fisheries
- Related to all of the above is interplay of public and private rights in water, which draws on aspects of eminent domain law and the federal commerce clause powers
- Water project law: the highly developed law regarding the formation, operation, and finance of public and quasi-public entities which operate local public works of flood control, navigation control, irrigation, and avoidance of environmental degradation
- Treaty Rights of Native Americans

The law governing these topics comes from all layers of law. Some derives from common law principles which have developed over centuries, and which evolve as the nature of disputes presented to courts change. For example, the judicial approach to landowner rights to divert surface waters has changed significantly in the last century as public attitudes about land and water have evolved. Some derives from state statutory law. Some derives from the original public grants of land to the States and from the documents of their origination. Some derives from state, federal and local regulation of waters through zoning, public health and other regulation. Non-federally recognized Indian tribes do not have water rights.

Water Law in the European Union

For countries within the European Union, water-related directives are important for water resource management and environmental and water quality standards.

Key directives include the Urban Waste Water Treatment Directive 1992 (requiring most towns and cities to treat their wastewater to specified standards), and the Water Framework Directive 2000, which requires water resource plans based on river basins, including public participation based on Aarhus Convention principles.

Clean Water Act

The Clean Water Act is the primary federal law in the United States governing water pollution. Commonly abbreviated as the CWA, the act established the goals of eliminating releases of high amounts of toxic substances into water, eliminating additional water pollution by 1985, and ensuring that surface waters would meet standards necessary for human sports and recreation by 1983.

The principal body of law currently in effect is based on the Federal Water Pollution Control Amendments of 1972 and was significantly expanded from the Federal Water Pollution Control Amendments of 1948. Major amendments were enacted in the Clean Water Act of 1977 and the Water Quality Act of 1987. The Clean Water Act does not directly address groundwater contamination. Groundwater protection provisions are included in the Safe Drinking Water Act, Resource Conservation and Recovery Act, and the Superfund act.

Clean Water Act (Ontario): The Clean Water Act (S.O. 2006, Chapter 22) is a law enacted by the Legislative Assembly of Ontario, Canada. The purpose of this Act is to protect existing and future sources of drinking water.

The *Clean Water Act, 2006* (Bill 43) is a major part of the Ontario government's commitment to ensuring that every Ontarian has access to safe drinking water. Protecting water at its source is the first step in the multi-barrier approach to source water protection. By stopping contaminants from getting into sources of drinking water — lakes, rivers and aquifers — we can provide the first line of defense in the protection of our environment and the health of Ontarians. For the first time, communities will be required to create and carry out a plan to protect the sources of their municipal drinking water supplies. The *Clean Water Act, 2006* will:

- Require local communities to look at the existing and potential threats to their water and set out and implement the actions necessary to reduce or eliminate significant threats.

- Empower communities to take action to prevent threats from becoming significant.
- Require public participation on every local source protection plan. This means everyone in the community gets a chance to contribute to the planning process.
- Require that all plans and actions are based on sound science.

This legislation sets out a basic framework for the establishment of community based groups that represent a cross section of sectors and geographic areas within Conservation Authority (Canada) boundaries. These community based groups are called Source Protection Committees and there are 19 such groups across Ontario. The *Clean Water Act, 2006* also introduced the Ontario Drinking Water Stewardship Program (ODWSP) a financial assistance program for farmers, landowners and small or medium businesses for activities that reduce threats to municipal drinking water sources.

Clean Water Protection Act

The Clean Water Protection Act (H.R. 1310) is a bill introduced in the 111th United States Congress via the House of Representatives Subcommittee on Water Resources and Environment, of the Committee on Transportation and Infrastructure. It proposes to redefine "fill material" to not include mining "waste" under the Federal Water Pollution Control Act.

It was introduced by Frank Pallone and Chris Shays in the 110th Congress, and almost identical bills had been introduced in previous years. As of July 2009, H.R.1310 had 154 co-sponsors. Several environmental groups such as Appalachian Voices, Kentuckians for the Commonwealth and Sierra Club, as well as many others, support the legislation because it will help to curb mountaintop removal mining. It would also affect situations similar to the Kensington Mine operation near Juneau, Alaska where a mine will be permitted to discharge mining waste into a small lake as "fill."

Colorado River Water Conservation District v. United States

Colorado River Water Conservation District v. United States, 424 U.S. 800 (1976), was a case in which the Supreme Court of the United States created a new doctrine of abstention, to prevent duplicative litigation between state and federal courts.

Background

In the Southwestern United States, water scarcity was (and remains) a critical problem. The McCarran Amendment, 43 U.S.C. § 666, was a statute enacted by United States Congress in 1952 allowing the United States to be joined as a defendant in certain suits concerning the adjudication or administration of rights to use of waters. In 1969, the state of Colorado enacted a law to attempt to reorganize the procedure for legal determination of water claims within the state.

The procedure that Colorado created divided the state into seven Water Divisions, each one encompassing one or more drainage basins for the largest rivers in the state. Each month, Water Referees in each division would rule on applications for water rights, or refer the case to a Water Judge, who would rule on referred or contested applications on a six-month schedule, applying the prior appropriation doctrine. A State Engineer, along with engineers for each division, were responsible for the administration and distribution of waters in each division.

Facts and Procedural History

The federal government reserves the water rights on federal lands such as Indian reservations and national parks and forests. The rights to such federal lands in the state of Colorado affect the rights of users in Colorado Water Division No.7. On November 14, 1972, the government filed suit in the United States District Court for the District of Colorado, asserting jurisdiction under both 28 U.S.C. § 1331 (the federal question jurisdiction statute) and 28 U.S.C. § 1345, which grants the United States district courts with original jurisdiction over all civil actions in which the United States is the plaintiff. The government sued for a declaratory judgment asserting its own reserved rights and those of several Native American tribes, against over 1,000 water users named as defendants. The government had previously filed similar suits in three other divisions.

Shortly after the suit over Division No.7 was commenced, one of the defendants in that suit filed an application in state court to join the United States as a party in a state court proceeding, pursuant to the McCarran Amendment. Several defendants in the federal court suit then moved to dismiss, challenging the federal courts jurisdiction to continue to hear matters pertaining to water rights. The district court granted the defendant's motion on ground of abstention doctrine, but the United States Court of Appeals for the Tenth Circuit reversed, holding that abstention was inappropriate.

Majority Opinion

Justice Brennan wrote for the majority. He first determined that the McCarran Amendment did not create an express or implied exception to § 1331 or § 1345, and then turned to the issue of whether or not the district court should have dismissed the suit because of the concurrent proceedings in state court. The McCarran Amendment permitted adjudication of the Native American tribes' rights in state court, and Brennan held there was no strong policy argument to eliminate the state courts' jurisdiction in this area. Moreover, the circumstances necessitating abstention were not present here, largely because this case did not present complex issues of state law or policy to be resolved, as was the case in past abstention cases such as *Railroad Commission v. Pullman Co.* (1941), *Burford v. Sun Oil Co.* (1943), or *Louisiana Power & Light Co. v. City of Thibodaux* (1959). Nor was this an attempt to enjoin state criminal proceedings, as in *Younger v. Harris* (1971).

However, Brennan also ruled that even though the case did not fall into any of the previously recognized categories of abstention, certain principles of judicial administration militated in favor of upholding the district court's decision to dismiss the case. Even though dismissal of a federal suit to avoid duplicative litigation was something of an exceptional nature, Brennan determined that it was permissible in some circumstances. He compared this case to an *in rem* action over the disposition of property, wherein a court first assuming jurisdiction may exercise jurisdiction to the exclusion of other courts. He listed factors federal courts may consider in determining the appropriateness of dismissal where concurrent jurisdiction exists:

- Inconvenience of the federal forum
- The desirability of avoiding piecemeal litigation
- The order in which jurisdiction was obtained by the concurrent forums

Brennan concluded by explaining that the whole policy advanced by the McCarran Amendment was to avoid "piecemeal adjudication of water rights," which would lead to concurrent litigation with inconsistent outcomes. Congress was well aware of pre-existing state procedures to determine water rights when it passed the McCarran Amendment.

Furthermore, the federal government was already involved in pending state lawsuits in three other Water Divisions (4, 5, and 6), there was a 300-mile distance between the federal courthouse in Denver and the state courthouse in Division 7 (which potentially created transportation difficulties for the over 1,000 defendants named in the federal suit), and there had been very few proceedings in the federal suit prior to the district court's dismissal. Brennan held that when all these factors were taken into account, the district court ruled correctly in dismissing the suit.

Dissenting Opinions

Justice Stewart's Dissent: Justice Stewart agreed with the majority that the McCarran Amendment did not diminish the federal courts' jurisdiction, and that the conventional doctrines of abstention were not implicated here. He disputed the majority's comparison of this suit to an *in rem* action, because a rule creating exclusive jurisdiction for the first court to take control of the property only applies when exclusive control of the property in question is required. In the litigation of water rights, where actual administration or control of a river was not being determined, this rule was clearly unnecessary.

Additionally, Stewart reasoned that dismissal of the federal suit was unnecessary because the rights the federal government sought to uphold were different from those being litigated in the state courts.

Specifically, the federal suit was about rights the federal government had previously reserved, rather than rights based on the prior appropriation doctrine. Finally, Stewart gave two more reasons not to dismiss the federal suit: issues of federal law were involved, as were the rights of Native American tribes (who were typically free of state jurisdiction).

Justice Stevens' dissent

Justice Stevens also added a brief dissenting opinion. He felt that the majority was unfairly shrinking the scope of federal jurisdiction, and that it was anomalous to forbid the federal government the right to a federal forum. He urged the affimance of the opinion of the Tenth Circuit.

Drainage Law

Drainage law is a specific area of water law related to drainage of surface water on real property. It is of great importance in areas where freshwater is scarce, where flooding is common, or where water is in high demand for agricultural or commercial purposes.

Drainage Law in the United States

In the United States, regulation of drainage is typically done on the state and local level. In addition to whatever statutes or local ordinances may be in effect in a given locality, there are three basic legal doctrines which the various state courts recognize.

In the state of Michigan, drainage law is so important that counties elect a drain commissioner to oversee water resources.

The Common Enemy Doctrine

The common enemy doctrine is a rule derived from English common law. It holds that since surface water is a "common enemy" to landowners, each landowner has the right to alter the drainage pattern of his land (for example by building dikes or drainage channels) without regard for the effects on neighboring parcels, as long as that water flows to where it otherwise would have naturally flowed.

Typically, a landowner can capture surface water (e.g. by rain barrels or dams) as well, and lower landowners will not have a cause of action unless the diversion is malicious. This rule is followed by approximately half the U.S. states, although some states have modified the doctrine to hold landowners liable for negligent damage to the parcels belonging to neighboring landowners.

The Civil Law Rule

The civil law rule (so named because it is derived from the civil law systems of France and Spain) is effectively the opposite of the common enemy doctrine. It holds that the owner of a lower parcel of land must accept the natural drainage from those parcels above his, and cannot alter the drainage pattern of his own land to increase the drainage flow onto parcels lower than his own. For this reason, this rule is sometimes referred to as the *"natural flow rule"*. Application of the civil law rule in its purest form would inhibit the development of land, since virtually every improvement on a parcel would alter the natural drainage. For this reason, this rule has been modified in those jurisdictions that use it, to permit reasonable changes in natural flow, often weighing the competing interests of neighboring landholders with the benefit of the development of the parcel.

The Reasonable Use Rule

The reasonable use rule presents an alternative to both the common enemy doctrine and the civil law rule. It allows a landowner to make "reasonable" alteration to the drainage pattern of his parcel, with liability only occurring when the alteration causes "unreasonable" harm toward neighboring parcels. Judicial mitigation of the common enemy doctrine and civil law rule often results in an approximation of the reasonable use rule.

Because the reasonable use rule presents a subjective standard, courts will often employ a balancing test to determine whether a landowner is liable to his neighbors for his alteration of drainage. For example, under the Restatement of Torts, the test was:

1. Was there reasonable necessity for the property owner to alter the drainage to make use of their land?
2. Was the alteration done in a reasonable manner?
3. Does the utility of the actor's conduct reasonably outweigh the gravity of harm to others?

Drought

A drought (or drouth [archaic]) is an extended period of months or years when a region notes a deficiency in its water supply. Generally, this occurs when a region receives consistently below average precipitation. It can have a substantial impact on the ecosystem and agriculture of the affected region. Although droughts can persist for several years, even a short, intense drought can cause significant damage and harm the local economy.

This global phenomenon has a widespread impact on agriculture. The United Nations estimates that an area of fertile soil the size of Ukraine is lost every year because of drought, deforestation, and climate instability. Lengthy periods of drought have long been a key trigger for mass migration and played a key role in a number of ongoing migrations and other humanitarian crises in the Horn of Africa and the Sahel.

Consequences

Periods of drought can have significant environmental, agricultural, health, economic and social consequences. The effect varies according to vulnerability. For example, subsistence farmers are more likely to migrate during drought because they do not have alternative food sources. Areas with populations that depend on as a major food source are more vulnerable to drought-triggered famine.

Drought can also reduce water quality, because lower water flows reduce dilution of pollutants and increase contamination of remaining water sources. Common consequences of drought include:

- Diminished crop growth or yield productions and carrying capacity for livestock
- Dust bowls, themselves a sign of erosion, which further erode the landscape
- Dust storms, when drought hits an area suffering from desertification and erosion
- Famine due to lack of water for irrigation
- Habitat damage, affecting both terrestrial and aquatic wildlife
- Malnutrition, dehydration and related diseases
- Mass migration, resulting in internal displacement and international refugees
- Reduced electricity production due to reduced water flow through hydroelectric dams
- Shortages of water for industrial users
- Snake migration and increases in snakebites
- Social unrest
- War over natural resources, including water and food
- Wildfires, such as Australian bushfires, are more common during times of drought

Globally

Drought is a normal, recurring feature of the climate in most parts of the world. It is among the earliest documented climatic events, present in the Epic of Gilgamesh and tied to the biblical story of Joseph's arrival in and the later

Exodus from Ancient Egypt. Hunter-gatherer migrations in 9,500 BC Chile have been linked to the phenomenon, as has the exodus of early man out of Africa and into the rest of the world around 135,000 years ago.

Modern peoples can effectively mitigate much of the impact of drought through irrigation and crop rotation. Failure to develop adequate drought mitigation strategies carries a grave human cost in the modern era, exacerbated by ever-increasing population densities.

Regions

Recurring droughts leading to desertification in the Horn of Africa have created grave ecological catastrophes, prompting massive food shortages, still recurring. To the north-west of the Horn, the Darfur conflict in neighboring Sudan, also affecting Chad, was fueled by decades of drought; combination of drought, desertification and overpopulation are among the causes of the Darfur conflict, because the Arab Baggara nomads searching for water have to take their livestock further south, to land mainly occupied by non-Arab farming peoples.

Approximately 2.4 billion people live in the drainage basin of the Himalayan rivers. India, China, Pakistan, Bangladesh, Nepal and Myanmar could experience floods followed by droughts in coming decades. Drought in India affecting the Ganges is of particular concern, as it provides drinking water and agricultural irrigation for more than 500 million people. The west coast of North America, which gets much of its water from glaciers in mountain ranges such as the Rocky Mountains and Sierra Nevada, also would be affected.

In 2005, parts of the Amazon basin experienced the worst drought in 100 years. A 23 July 2006 article reported Woods Hole Research Center results showing that the forest in its present form could survive only three years of drought. Scientists at the Brazilian National Institute of Amazonian Research argue in the article that this drought response, coupled with the effects of deforestation on regional climate, are pushing the rainforest towards a "tipping point" where it would irreversibly start to die. It concludes that the rainforest is on the brink of being turned into savanna or desert, with catastrophic consequences for the world's climate. According to the WWF, the combination of climate change and deforestation increases the drying effect of dead trees that fuels forest fires.

By far the largest part of Australia is desert or semi-arid lands commonly known as the outback. A 2005 study by Australian and American researchers investigated the desertification of the interior, and suggested that one explanation was related to human settlers who arrived about 50,000 years ago. Regular burning by these settlers could have prevented monsoons from reaching interior Australia. In June 2008 it became known that an expert panel had warned of long term, maybe irreversible, severe ecological damage for the whole Murray-Darling basin if it does not receive sufficient water by October. Australia could experience more severe droughts and they could become more frequent in the future, a government-commissioned report said on July 6, 2008. The Australian of the year 2007, environmentalist Tim Flannery, predicted that unless it made drastic changes, Perth in Western Australia could become the world's first ghost metropolis, an abandoned city with no more water to sustain its population.

East Africa currently faces its worst drought in decades, with crops and livestock destroyed. The U.N. World Food Programme recently said that nearly four million Kenyans urgently needed food.

Causes

Ancient Meso-American civilizations of the Mayans and Aztecs likely amplified droughts in the Yucatan and southern Mexico by clearing rainforests to make room for pastures and farmland.

Generally, rainfall is related to the amount of water vapor in the atmosphere, combined with the upward forcing of the air mass containing that water vapor. If either of these are reduced, the result is a drought. This can be triggered by an above average prevalence of high pressure systems, winds carrying continental, rather than oceanic air masses (i.e. reduced water content), and ridges of high pressure areas form with behaviors which prevent or restrict the developing of thunderstorm activity or rainfall over one certain region. Oceanic and atmospheric weather cycles such as the El Nipo-Southern Oscillation (ENSO) make drought a regular recurring feature of the Americas along the Midwest and Australia. *Guns, Germs, and Steel* author Jared Diamond sees the stark impact of the multi-year ENSO cycles on Australian weather patterns as a key reason that Australian aborigines remained a hunter-gatherer society rather than adopting agriculture. Another climate oscillation known as the North Atlantic Oscillation has been tied to droughts in northeast Spain.

Human activity can directly trigger exacerbating factors such as over farming, excessive irrigation, deforestation, and erosion adversely impact the ability of the land to capture and hold water. While these tend to be relatively isolated in their scope, activities resulting in global climate change are expected to trigger droughts with a substantial impact on agriculture throughout the world, and especially in developing nations. Overall, global warming will result in increased world rainfall. Along with drought in some areas, flooding and erosion will increase in others. Paradoxically, some proposed solutions to global warming that focus on more active techniques, solar radiation management through the use of a space sunshade for one, may also carry with them increased chances of drought.

Types of Drought

As a drought persists, the conditions surrounding it gradually worsen and its impact on the local population gradually increases. People tend to define droughts in three main ways:

1. Meteorological drought is brought about when there is a prolonged period with less than average precipitation. Meteorological drought usually precedes the other kinds of drought.
2. Agricultural droughts are droughts that affect crop production or the ecology of the range. This condition can also arise independently from any change in precipitation levels when soil conditions and erosion triggered by poorly planned agricultural endeavors cause a shortfall in water available to the crops. However, in a traditional drought, it is caused by an extended period of below average precipitation.
3. Hydrological drought is brought about when the water reserves available in sources such as aquifers, lakes and reservoirs fall below the statistical average. Hydrological drought tends to show up more slowly because it involves stored water that is used but not replenished. Like an agricultural drought, this can be triggered by more than just a loss of rainfall. For instance, Kazakhstan was recently awarded a large amount of money by the World Bank to restore water that had been diverted to other nations from the Aral Sea under Soviet rule. Similar circumstances also place their largest lake, Balkhash, at risk of completely drying out.

Drought Protection and Relief

Strategies for drought protection, mitigation or relief include:

- Dams - many dams and their associated reservoirs supply additional water in times of drought.
- Cloud seeding - an artificial technique to induce rainfall.
- Desalination of sea water for irrigation or consumption.
- Drought monitoring - Continuous observation of rainfall levels and comparisons with current usage levels can help prevent man-made drought. For instance, analysis of water usage in Yemen has revealed that their water table (underground water level) is put at grave risk by over-use to fertilize their Khat crop. Careful monitoring of moisture levels can also help predict increased risk for wildfires, using such metrics as the Keetch-Byram Drought Index or Palmer Drought Index.
- Land use - Carefully planned crop rotation can help to minimize erosion and allow farmers to plant less water-dependent crops in drier years.

- Outdoor water-use restriction - Regulating the use of sprinklers, hoses or buckets on outdoor plants, filling pools, and other water-intensive home maintenance tasks.
- Rainwater harvesting - Collection and storage of rainwater from roofs or other suitable catchments.
- Recycled water - Former wastewater (sewage) that has been treated and purified for reuse.
- Transvasement - Building canals or redirecting rivers as massive attempts at irrigation in drought-prone areas.

Water Right

Water right in water law refers to the right of a user to use water from a water source, e.g., a river, stream, pond or source of groundwater. In areas with plentiful water and few users, such systems are generally not complicated or contentious. In other areas, especially arid areas where irrigation is practiced, such systems are often the source of conflict, both legal and physical. Some systems treat surface water and ground water in the same manner, while others use different principles for each.

Types of Water Rights

There are fundamental differences between the nature and source of water rights in different countries, including that they can be land-based rights, use-based rights, or rights based on ownership of water bodies. In addition, some countries apply the principle of water solidarity, which has become increasingly important in the EU. This principle attempts to reconcile water rights through compromise solutions based on the public interest, and it has been written into legislation in France, for example.

Land-based or Riparian Rights

Riparian rights are based on land ownership, and are protected by property law. Riparian rights state that only the owner of the banks of the water source have a right to the 'undiminished, unaltered flow' of the water. Riparian rights are only transferable when the riparian land ownership title is transferred to a new owner.

Use-based Rights

Use-based water rights are protected by the law of torts. Use-based rights state that land ownership is not essential, as long as water users have legal access to the water source. There is a hierarchy of use, where the first user has the strongest rights (first in, first served). Rights users can only enforce rights against users with lower ranks (those who came later). Use-based rights are usufructuary, fully transferable to anyone.

Water Rights based on Ownership of Water Bodies

In Finland, for example, waterbodies are generally privately owned, which is not the case in most EU countries, but Finland also applies the Roman law principle of *aqua profluens* (flowing water), according to which the freely flowing water in waterbodies cannot be owned or possessed. This means that the owners of waterbodies cannot prohibit diversion of water for agricultural, industrial, municipal, or domestic use according to the provisions of the Finnish Water Law and cannot prohibit use of the waterbodies for recreational purposes.

History of Water Rights

In Roman times, the law was that people could obtain temporary usufructuary rights for running water. These rights were independent of land ownership, and lasted as long as use continued. Under Roman law, no land was "owned" by citizens, it was all owned by the "republic" and controlled by politicians. In Medieval times, the common law of the day treated an all freshwater streams as static, meaning landowners owned parts of rivers, with full accompanying rights. Landowners could also seek damages for loss of water diverted upstream. Non-landowners did not have use rights, except by obtaining a prescription.

Over time, rights evolved from being land based to use based, allowing non-landowners to hold enforceable rights. A reasonable use theory evolved in some countries.

Water Rights in the United States

In the United States, there are two divergent systems for determining water rights. Riparian water rights (derived from English common law) are common in the east and prior appropriation water rights (developed in Colorado and California) are common in the west. Each state has its own variations on these basic principles, as informed by custom, culture, geography, legislation and case law. California law, for example, includes elements of both systems. In general, a water right is established by obtaining an authorization from the state in the form of a water right permit. A legal right is formally consummated, or perfected, by exercising the water right permit and using the water for a beneficial purpose. Under the prior appropriation doctrine, water rights are "first in time, first in right." That is, the older, or senior, water right may operate to the exclusion of junior water rights. The concept of "priority date" is significant. The priority date is generally associated with the date that water was first put to beneficial use, or the date that a successful application for a water right was submitted, and indicates the relative status of seniority among competing users. Older rights are senior. More recent rights are junior. Water rights are generally established pursuant to State law, but there are exceptions, most notably, the concept of federal reserved water rights. Reserved water rights are rights that are established when the federal government reserves land for a specific federal purpose. Courts have held that there is an implied water right to satisfy the primary purposes of the reservation. Examples of reservations include Indian reservations, national wildlife refuges, federal forests and military bases. Proceedings to determine the relative priority of claims to water rights are known as adjudications. Through Congress's passage of the McCarren amendment, the federal government has consented to having its claims adjudicated in state courts. All states offer mechanisms for changing how a water right is exercised, e.g., amending the point of diversion or withdrawal, the place of use and the purpose of use. In reviewing such requests, the state must guard against the impairment of other water rights, the enlargement of the water right and injury to the public interest.

Resolution of Interstate Water Conflicts

Because water bodies may cross political and jurisdictional boundaries, conflicts may arise. In the United States, three basic approaches are used to settle such conflicts: 1) Litigation before the Supreme Court of the United States; 2) Legislative resolution by the Congress of the United States; and 3) Negotiation and ratification of interstate compacts between states. In the western United States, for example, the 1922 Colorado River Compact divides the Colorado River basin into two areas, the Upper Division (comprising Colorado, New Mexico, Utah and Wyoming) and the Lower Division (Nevada, Arizona and California). A comprehensive review of existing interstate water compacts has been published by the Model Interstate Water Compact Project at the University of New Mexico School of Law's Utton Transboundary Resource Center.

Limitations on Water Rights

In California, courts have held that appropriation water rights may be limited under the public trust doctrine, a common law principle recognized by some courts, which holds that the public has access rights upon navigable waters and that navigable waters are held in trust for the use of the people. The public trust doctrine was invoked by the California Supreme Court in a case restricting the amount of water Los Angeles could divert from tributaries of Mono Lake. The case was filed by the Audubon Society and the Mono Lake Committee.

In the United States, Navigable waters are subject to the commerce clause of the U.S. Constitution. The commerce clause provides the federal government the ability to restrict state issued water rights via, for example, the enforcement of water quality standards via the Federal Water Pollution Control Act (Clean Water Act), the Federal Power Act and the protection of endangered species via the Endangered Species Act.

Water Crisis

Water crisis is a general term used to describe a situation where the available water within a region is less than the region's demand. The term has been used to describe the availability of potable water in a variety of regions by the United Nations and other world organizations. Others, for example the Food and Agriculture Organization, said in 2003 that there is no water crisis but steps must be taken to avoid one in the future. The major aspects of the water crisis are allegedly overall scarcity of usable water and water pollution.

Research

According to *Nature* (2010), about 80% of the world's population (5.6 billion in 2011) live in areas with threats to water security. The water security is a shared threat to human and nature and it is pandemic. Human water-management strategies can affect detrimentally to wildlife, such as migrating fish. Regions with intensive agriculture and dense population, as the US and Europe, have high threat to water security. The researcher estimate that during 2010-2015, ca US$800 billion will be required to cover the annual global investment in water infrastructure. Good management of water resources can jointly manage biodiversity protection and human water security. Preserving flood plains rather than constructing flood-control reservoirs would provide a cost-effective way to control floods while protecting the biodiversity of wildlife that occupies such areas.

The New York Times article, "Southeast Drought Study Ties Water Shortage to Population, Not Global Warming", summarizes the findings of Columbia University researcher on the subject of the droughts in the southwest between 2005 and 2007. The findings were published in the Journal of Climate. They say the water shortages resulted from population size more than rainfall. Census figures show that Georgia's population rose to 9.54 million from 6.48 million between 1990 and 2007. After studying data from weather instruments, computer models and measurements of tree rings which reflect rainfall, they found that the droughts were not unprecedented and result from normal climate patterns and random weather events. "Similar droughts unfolded over the last thousand years, the researchers wrote. Regardless of climate change, they added, similar weather patterns can be expected regularly in the future, with similar results." As the temperature increases, rainfall in the Southeast will increase but because of evaporation the area may get even drier. The researchers concluded with a statement saying that any rainfall comes from complicated internal processes in the atmosphere that are very hard to predict because of the large amount of variables.

Lawrence Smith, the president of the population institute, asserts that although an overwhelming majority of the planet is composed of water, 97%

of this water is constituted of saltwater; the fresh water used to sustain humans is only 3% of the total amount of water on Earth (Hoevel). Therefore, Smith believes that the competition for water in an overpopulated world would pose a major threat to human stability (Hoevel); indeed, world wars may be fought over the control of thinning ice sheets and nearly desiccated reservoirs. 2 billion people have gained access to a safe water source since 1990. The proportion of people in developing countries with access to safe water is calculated to have improved from 30 percent in 1970 to 71 percent in 1990, 79 percent in 2000 and 84 percent in 2004, parallel with rising population. This trend is projected to continue.

The Earth has a limited supply of fresh water, stored in aquifers, surface waters and the atmosphere. Sometimes oceans are mistaken for available water, but the amount of energy needed to convert saline water to potable water is prohibitive today, explaining why only a very small fraction of the world's water supply derives from desalination.

Manifestations

There are several principal manifestations of the water crisis.

- Inadequate access to safe drinking water for about 884 million people
- Inadequate access to water for sanitation and waste disposal for 2.5 billion people
- Groundwater overdrafting (excessive use) leading to diminished agricultural yields
- Overuse and pollution of water resources harming biodiversity
- Regional conflicts over scarce water resources sometimes resulting in warfare

Waterborne diseases and the absence of sanitary domestic water are one of the leading causes of death worldwide. For children under age five, waterborne diseases are the leading cause of death. At any given time, half of the world's hospital beds are occupied by patients suffering from waterborne diseases. According to the World Bank, 88 percent of all waterborne diseases are caused by unsafe drinking water, inadequate sanitation and poor hygiene. Water is the underlying tenuous balance of safe water supply, but controllable factors such as the management and distribution of the water supply itself contribute to further scarcity. A 2006 United Nations report focuses on issues of governance as the core of the water crisis, saying "There is enough water for everyone" and "Water insufficiency is often due to mismanagement, corruption, lack of appropriate institutions, bureaucratic inertia and a shortage of investment in both human capacity and physical infrastructure". Official data also shows a clear correlation between access to safe water and GDP per capita. It has also been claimed, primarily by economists, that the water situation has occurred because of a lack of property rights, government regulations and subsidies in the water sector, causing prices to be too low and consumption too high. Vegetation and wildlife are fundamentally dependent upon adequate freshwater resources. Marshes, bogs and riparian zones are more obviously dependent upon sustainable water supply, but forests and other upland ecosystems are equally at risk of significant productivity changes as water availability is diminished. In the case of wetlands, considerable area has been simply taken from wildlife use to feed and house the expanding human population. But other areas have suffered reduced productivity from gradual diminishing of freshwater inflow, as upstream sources are diverted for human use. In seven states of the U.S. over 80 percent of all historic wetlands were filled by the 1980s, when Congress acted to create a "no net loss" of wetlands. In Europe extensive loss of wetlands has also occurred with resulting loss of biodiversity. For example many bogs in Scotland have been developed or diminished through human population expansion. One example is the Portlethen Moss in Aberdeenshire.

On Madagascar's highland plateau, a massive transformation occurred that eliminated virtually all the heavily forested vegetation in the period 1970 to 2000. The slash and burn agriculture eliminated about ten percent of the total country's native biomass and converted it to a barren wasteland. These effects were from overpopulation and the necessity to feed poor indigenous peoples, but the adverse effects included widespread gully erosion that in turn produced heavily silted rivers that "run red" decades after the deforestation. This eliminated a large amount of usable fresh water and also destroyed much of the riverine ecosystems of several large west-flowing rivers. Several fish species have been driven to the edge of extinction and some, such as, the disturbed Tokios, coral reef formations in the Indian Ocean are effectively lost. In October 2008, Peter Brabeck-Letmathe, chairman and former chief executive of Nestlé, warned that the production of biofuels will further deplete the world's water supply.

Overview of Regions Suffering Crisis Impacts

There are many other countries of the world that are severely impacted with regard to human health and inadequate drinking water. The following is a partial list of some of the countries with significant populations (numerical population of affected population listed) whose only consumption is of contaminated water:

- Sudan 12.3 million
- Venezuela 5.0 million
- Ethiopia 2.7 million
- Tunisia 2.1 million
- Cuba 1.3 million

Several world maps showing various aspects of the problem can be found ·in this graph article.

According to the California Department of Water Resources, if more supplies aren't found by 2020, the region will face a shortfall nearly as great as the amount consumed today. Los Angeles is a coastal desert able to support at most 1 million people on its own water; the Los Angeles basin now is the core of a megacity that spans 220 miles (350 km) from Santa Barbara to the Mexican border. The region's population is expected to reach 41 million by 2020, up from 28 million in 2009. The population of California continues to grow by more than two million a year and is expected to reach 75 million in 2030, up from 49 million in 2009. But water shortage is likely to surface well before then.

Water deficits, which are already spurring heavy grain imports in numerous smaller countries, may soon do the same in larger countries, such as China and India. The water tables are falling in scores of countries (including Northern China, the US, and India) due to widespread overpumping using powerful diesel and electric pumps. Other countries affected include Pakistan, Iran, and Mexico. This will eventually lead to water scarcity and cutbacks in grain harvest. Even with the overpumping of its aquifers, China is developing a grain deficit. When this happens, it will almost certainly drive grain prices upward. Most of the 3 billion people projected to be added worldwide by mid-century will be born in countries already experiencing water shortages. Unless population growth can be slowed quickly it is feared that there may not be a practical non-violent or humane solution to the emerging world water shortage.

After China and India, there is a second tier of smaller countries with large water deficits — Algeria, Egypt, Iran, Mexico, and Pakistan. Four of these already import a large share of their grain. But with a population expanding by 4 million a year, it will also likely soon turn to the world market for grain.

According to a UN climate report, the Himalayan glaciers that are the sources of Asia's biggest rivers - Ganges, Indus, Brahmaputra, Yangtze, Mekong, Salween and Yellow - could disappear by 2035 as temperatures rise. It was later revealed that the source used by the UN climate report actually stated 2350, not 2035. Approximately 2.4 billion people live in the drainage basin of the Himalayan rivers. India, China, Pakistan, Bangladesh, Nepal and Myanmar could experience floods followed by droughts in coming decades. In India alone, the Ganges provides water for drinking and farming for more than 500 million people. The west coast of North America, which gets much of its water from glaciers in mountain ranges such as the Rocky Mountains and Sierra Nevada, also would be affected.

By far the largest part of Australia is desert or semi-arid lands commonly known as the outback. In June 2008 it became known that an expert panel had warned of long term, possibly irreversible, severe ecological damage for the whole Murray-Darling basin if it does not receive sufficient water by October. Water restrictions are currently in place in many regions and cities of Australia in response to chronic shortages resulting from drought. The Australian of the year 2007, environmentalist Tim Flannery, predicted that unless it made drastic changes, Perth in Western Australia could become the world's first ghost metropolis, an abandoned city with no more water to sustain its population. However, Western Australia's dams reached 50% capacity for the first time since 2000 as of September 2009. As a result, heavy rains have brought forth positive results for the region. Nonetheless, the following year, 2010, Perth suffered its second-driest winter on record and the water corporation tightened water restrictions for spring.

Effects on Climate

Aquifer drawdown or overdrafting and the pumping of fossil water increases the total amount of water within the hydrosphere subject to transpiration and evaporation processes, thereby causing accretion in water vapour and cloud cover, the primary absorbers of infrared radiation in the earth's atmosphere. Adding water to the system has a forcing effect on the whole earth system, an accurate estimate of which hydrogeological fact is yet to be quantified.

Outlook

Figure: *Wind and solar power such as this installation in a village in northwest Madagascar can make a difference in safe water supply.*

Construction of wastewater treatment plants and reduction of groundwater overdrafting appear to be obvious solutions to the worldwide problem; however, a deeper look reveals more fundamental issues in play. Wastewater treatment is highly capital intensive, restricting access to this technology in some regions; furthermore the rapid increase in population of many countries makes this a race that is difficult to win. As if those factors are not daunting enough, one must consider the enormous costs and skill sets involved to maintain wastewater treatment plants even if they are successfully developed.

Reduction in groundwater overdrafting is usually politically very unpopular and has major economic impacts to farmers; moreover, this strategy will necessarily reduce crop output, which is something the world can ill-afford, given the population level at present.

At more realistic levels, developing countries can strive to achieve primary wastewater treatment or secure septic systems, and carefully analyse wastewater outfall design to minimise impacts to drinking water and to ecosystems. Developed countries can not only share technology better, including cost-effective wastewater and water treatment systems but also in hydrological transport modeling. At the individual level, people in developed countries can look inward and reduce overconsumption, which further strains worldwide water consumption. Both developed and developing countries can increase protection of ecosystems, especially wetlands and riparian zones. These measures will not only conserve biota, but also render more effective the natural water cycle flushing and transport that make water systems more healthy for humans.

A range of local, low-tech solutions are being pursued by a number of companies. These efforts center around the use of solar power to distill water at temperatures slightly beneath that at which water boils. By developing the capability to purify any available water source, local business models could be built around the new technologies, accelerating their uptake.

Conventional Fossil or Nuclear Energy Based Desalination

As new technological innovations continue to reduce the capital cost of desalination, more countries are building desalination plants as a small element in addressing their water crises.

- Israel desalinizes water for a cost of 53 cents per cubic meter
- Singapore desalinizes water for 49 cents per cubic meter and also treats sewage with reverse osmosis for industrial and potable use (NEWater).
- China and India, the world's two most populous countries, are turning to desalination to provide a small part of their water needs
- In 2007 Pakistan announced plans to use desalination
- All Australian capital cities (except Darwin, Northern Territory and Hobart) are either in the process of building desalination plants, or are already using them. In late 2011, Melbourne will begin using Australia's largest desalination plant, the Wonthaggi desalination plant to raise low reservoir levels.
- In 2007 Bermuda signed a contract to purchase a desalination plant
- The largest desalination plant in the United States is the one at Tampa Bay, Florida, which began desalinizing 25 million gallons (95000 m^3)

of water per day in December 2007. In the United States, the cost of desalination is $3.06 for 1,000 gallons, or 81 cents per cubic meter. In the United States, California, Arizona, Texas, and Florida use desalination for a very small part of their water supply.

- After being desalinized at Jubail, Saudi Arabia, water is pumped 200 miles (320 km) inland though a pipeline to the capital city of Riyadh.

A January 17, 2008, article in the *Wall Street Journal* states, "World-wide, 13,080 desalination plants produce more than 12 billion gallons of water a day, according to the International Desalination Association." The world's largest desalination plant is the Jebel Ali Desalination Plant (Phase 2) in the United Arab Emirates. It is a dual-purpose facility that uses multi-stage flash distillation and is capable of producing 300 million cubic meters of water per year. A typical aircraft carrier in the U.S. military uses nuclear power to desalinize 400,000 US gallons (1,500,000 L) of water per day. While desalinizing 1,000 US gallons (3,800 L) of water can cost as much as $3, the same amount of bottled water costs $7,945. However, given the energy intensive nature of desalination, with associated economic and environmental costs, desalination is generally considered a last resort after water conservation. But this is changing as prices continue to fall. According to MSNBC, a report by Lux Research estimated that the worldwide desalinated water supply will triple between 2008 and 2020. However, not everyone is convinced that desalination is or will be economically viable or environmentally sustainable for the foreseeable future. Debbie Cook, the former mayor of Huntington Beach, California, has been a frequent critic of desalination proposals ever since she was appointed as a member of the California Desalination Task Force. Cook claims that in addition to being energy intensive, desalination schemes are very costly—often much more costly than desalination proponents claim. In her writing on the subject, Cook points to a long list of projects that have stalled or been aborted for financial or other reasons, and suggests that water-stressed regions would do better to focus on conservation or other water supply solutions than to invest in desalination plants.

Solar Energy Based Desalination

A novel approach to desalination is the Seawater Greenhouse which takes seawater and uses solar energy to desalinate it in conjunction with growing food crops in a specially adapted greenhouse.

Global experiences in Managing Water Crisis

It is alleged that the likelihood of conflict rises if the rate of change within the basin exceeds the capacity of institution to absorb that change. Although water crisis is closely related to regional tensions, history showed that acute conflicts over water are far less than the record of cooperation.

The key lies in strong institutions and cooperation. The Indus River Commission and the Indus Water Treaty survived two wars between India and Pakistan despite their hostility, proving to be a successful mechanism in resolving conflicts by providing a framework for consultation inspection and exchange of data. The Mekong Committee has also functioned since 1957 and survived the Vietnam War. In contrast, regional instability results when there is an absence of institutions to co-operate in regional collaboration, like Egypt's plan for a high dam on the Nile. However, there is currently no global institution in place for the management and management of trans-boundary water sources, and international co-operation has happened through ad hoc collaborations between agencies, like the Mekong Committee which was formed due to an alliance between UNICEF and the US Bureau of Reclamation. Formation of strong international institutions seems to be a way forward - they fuel early intervention and management, preventing the costly dispute resolution process.

One common feature of almost all resolved disputes is that the negotiations had a "need-based" instead of a "right–based" paradigm. Irrigable lands, population, technicalities of projects define "needs". The success of a need-based paradigm is reflected in the only water agreement ever negotiated in the Jordan River Basin, which focuses in needs not on rights of riparians. In the Indian subcontinent, irrigation requirements of Bangladesh determine water allocations of The Ganges River. A need based, regional approach focuses on satisfying individuals with their need of water, ensuring that minimum quantitative needs are being met. It removes the conflict that arises when countries view the treaty from a national interest point of view, move away from the zero-sum approach to a positive sum, integrative approach that equitably allocated the water and its benefits.

In Culture

Religion

Water is considered a purifier in most religions. Major faiths that incorporate ritual washing (ablution) include Christianity, Islam, Hinduism, Rastafari movement, Shinto, Taoism, Judaism, and Wicca. Immersion (or aspersion or affusion) of a person in water is a central sacrament of Christianity (where it is called baptism); it is also a part of the practice of other religions, including Judaism (*mikvah*) and Sikhism (*Amrit Sanskar*). In addition, a ritual bath in pure water is performed for the dead in many religions including Judaism and Islam. In Islam, the five daily prayers can be done in most cases after completing washing certain parts of the body using clean water (*wudu*). In Shinto, water is used in almost all rituals to cleanse a person or an area (e.g., in the ritual of *misogi*). Water is mentioned numerous times in the Bible, for example: "The earth was formed out of water and by water" (NIV). In the Qur'an it is stated that "Living things are made of water" and it is often used to describe paradise.

Philosophy

The Ancient Greek philosopher Empedocles held that water is one of the four classical elements along with fire, earth and air, and was regarded as the ylem, or basic substance of the universe. Water was considered cold and moist. In the theory of the four bodily humors, water was associated with phlegm. The classical element of Water was also one of the five elements in traditional Chinese philosophy, along with earth, fire, wood, and metal.

Water is also taken as a role model in some parts of traditional and popular Asian philosophy. James Legge's 1891 translation of the Dao De Jing states "The highest excellence is like (that of) water. The excellence of water appears in its benefiting all things, and in its occupying, without striving (to the contrary), the low place which all men dislike. Hence (its way) is near to (that of) the Tao" and "There is nothing in the world more soft and weak than water, and yet for attacking things that are firm and strong there is nothing that can take precedence of it—for there is nothing (so effectual) for which it can be changed."

Literature

Water is used in literature as a symbol of purification. Examples include the critical importance of a river in *As I Lay Dying* by William Faulkner and the drowning of Ophelia in *Hamlet*.

Sherlock Holmes held that "From a drop of water, a logician could infer the possibility of an Atlantic or a Niagara without having seen or heard of one or the other."

4

Water Sources and Water Availability

In each irrigation area there is a demand for water and a supply of water. The demand for water varies over time and depends on the types of crops, crop growth stages and on the climate. While transporting irrigation water from the water source and applying it to the plant roots, a portion of the water is lost through evaporation, leakage from the canals and percolation below the roots of the crop. The scheme's gross irrigation need, SIN_{gross} includes these water losses. To express the percentage of irrigation water that is used efficiently and what percentage is lost, the term irrigation efficiency is used.

The supply of water to the irrigation system, SWS, should be almost if not equal to the demand for water, SIN_{gross}, which is the total of the irrigation need for all fields plus the water loss during transport and distribution.

When the supply of water exceeds the demand, plants may receive too much water which has a negative effect on their growth. Or, on the other hand, costly water may be spilled and disappear into the drainage system. When the supply is less than the demand, the irrigation area may suffer from drought and plant production will decrease.

Extension agents and field technicians need to be able to advise farmers on how to handle existing as well as anticipate future problems of water shortage. Problems may arise when an irrigation area is extended, or when new crops are introduced. The technician therefore needs to have an understanding of water shortage problems together with a knowledge of the correct approach, albeit simplified, to matching the supply with the demand for water. This chapter deals with the supply aspects of irrigation. Various types of water sources are presented which are largely what determine water availability. River and ground-water hydrology are beyond the scope of this manual, so that only a conceptual example is provided.

Water Sources

The water needed to supply an irrigation scheme is taken from a water source. The most common sources of water for irrigation include rivers, reservoirs and lakes, and ground-water.

Water Availability

The possibility of supplying as much water to the irrigation area as is needed during each period of the irrigation season depends primarily on the availability of the water at its source. Availability may vary a lot over the year, or even between one year and another. Secondly, the supply depends on the capacity of the facility installed to withdraw the water from the water source.

Further, technicians should be aware that water must be available during each week or month of the growing season. This is illustrated with an example.

It is thus important to know how much water one can tap from a given water source over the course of a season or a year, when:

- developing a new irrigation scheme,
- extending an existing scheme,
- changing the cropping season,
- adding a second or third irrigation season,
- switching from a low water demand crop to a crop such as rice with high water requirements.

The scheme irrigation water supply depends, in technical terms, on two factors:

- whether the water source itself has a limited flow or a limited volume,
- whether the facilities that control the tapping (gate, well, pump) and conveyance (canal, pipe) have a limited capacity.

There may also exist water regulations which prohibit an unlimited withdrawal of water from the source.

Methods of Tapping Water from Water Sources

Water for irrigation has to be carried from the river, the reservoir, the lake or the ground-water to the field. This can be done in two different ways:

- by making use of gravity. A simple gravity system can only be used if the water level of the river or reservoir is higher than the level of the fields in the irrigation scheme;

- by using a pump to lift the water above ground level and then let it flow to the fields.

The tapping of water from a river or reservoir by gravity, and the pumping of water from a river, lake or ground-water are explained briefly.

Rivers

Rivers are used all over the world as sources of irrigation water. The most typical quality defining a river is that it flows, it is not a reservoir which contains a fixed amount of water. At each moment a new amount of water is passing any given location along the river. The flow of river fluctuates over time. The flows of some rivers fluctuate greatly over relatively short periods of time; these are mainly small local rivers which respond quickly to rainfall in their catchment area.

Other rivers show little fluctuation or vary only over a long period of time. These are mainly rivers with a large catchment area, where the rains are spread over a greater area and for a longer period of time. River flows vary considerably, not only within a given year, but also from one year to the next. In a year with little rain during the rainy season, the river flow will be small; sometimes the river flow will cease altogether during the dry season. The river flow will be far more in years with heavy rainfall during the wet season.

The discharge of rivers is commonly expressed in cubic metres per second (m^3/s). A simple method of measuring the discharge in canals is described in Training Manual 7, entitled "Canals". Although the method is applied to canals, it can also be used for small rivers.

Tapping Water from Rivers

Direct river diversion: The offtake canal is directly excavated through the river bank and no gate is provided. Consequently, there is no control of the discharge into the canal. The diverted discharge depends on the water level in the river. That means the discharge will be high during periods of high water and low during periods with a low water level.

Usually, to avoid changes in the discharge, a control gate can be installed. The gate can then be opened completely when the river level is low, and opened only partly when the water level is high.

The advantage of this system of water abstraction is its relatively low cost. However, the method can only be used when the river water level, even at low flows, is higher than the land which is to be irrigated. In other words, the point of diversion must be located at some distance upstream of the irrigation area, where the water level in the river is still higher than the level of the irrigation area. If the area is very flat, the canal distance may be too long to justify this simple river diversion system.

River Diversion using a Weir: To avoid the problems caused by fluctuating water levels in a river, a weir can be built across the river. During periods of high river discharges, water will flow over the weir. The water level upstream of the weir will show little variation during the year, and it will remain higher during the dry season than it would without the weir.

The water level upstream is raised by some 20 cm and the flow is diverted to an irrigated plot through an offtake on the right bank. A low section of the structure functions as a spillway and discharges the surplus water downstream.

Pumping from a River: The cost of pumping water increases with the size of the scheme. The cost of a weir, however, depends largely on the size of the river and only partly on the size of the scheme. Therefore, pumping is often used to irrigate small areas whose sources of water are larger rivers. Also, this may be a much less expensive solution than building a weir. Pumping is essential if irrigation by gravity is not possible. This will be the case when rivers lie well below field level.

Availability of River Flow

It is important to make sure that there will be no lack of irrigation water. If water is in short supply during some part of the irrigation season, crop production will suffer, returns will decline and part of the scheme's investment will lay idle.

Irrigation schemes depend on the supply of water from the river throughout the irrigation season. The reliability of that supply is an essential factor in the success of the irrigation scheme. Thus, once the SIN_{gross} of a scheme has been estimated, the flow of the river and the reliability of that flow during the irrigation season must be ascertained.

The availability of river water will fall into one of the three scenarios described below.

- The river flow is sufficient to meet the SIN_{gross} only when the flow is abundant. Because the river flow fluctuates during the year, and over the years, water shortages do occur when the flow is low during dry months of the year and during dry years. It may be ascertained during which months of the year the river has its lowest discharge. It is clear that under such circumstances any change in irrigation practices resulting in higher irrigation needs during the dry months should be discouraged.
- The river flow is abundant as compared to the SIN_{gross} even during dry months of the irrigation season, but water shortages do occur in dry years. A change in irrigation practices which requires more water, or small scheme extensions, may result in more severe water shortages in the future.

- The river flow is always much greater than the SIN_{gross} over the irrigation season, and this remains true over the years. There has thus never been a problem for the irrigation scheme as regards water availability. This can be checked with local farmers. In this case, there will be no problem with small scheme extensions, or with the introduction of crops, such as rice, which consume a lot of water.

The best solution is to consult the hydrological service. This service has information on water resources and river flows as they collect hydrological data, such as rainfall and discharges of rivers and streams. They can estimate the minimum river flow during the irrigation season based on historical records and predict the minimum river flow for a dry year. If, however, no measurements of river discharge are available, then data collection must start as soon as possible so that future plans may be made based on reliable data.

It may happen that the river flow during one part of the irrigation season is limited, while at other times during the season a large surplus of water is available. Part of this surplus can be stored by constructing a dam in the river. The stored water can then be used in addition to the available river flow. The cost of building a dam however is high, and expert help should be sought to study alternatives and produce appropriate designs.

Reservoirs and Lakes

Lakes are natural depressions of the land which are filled up with water. Fresh water lakes have a natural outlet through which the lake discharges superfluous water. Lakes are supplied with water by rainfall that falls directly on the surface of the lake, by water run-off from the adjacent land and small streams, or by ground-water that seeps through the soil to the lowest point which is the lake. Lakes lose water via evaporation from the lake's surface, via the lake's natural outlet (overflow), or through percolation from the bottom of the lake to the ground-water. A reservoir is an artificial lake. It can be formed by building a dam across a valley, by excavating the land or by surrounding a piece of land with dykes. The water is stored in the reservoir and can be used for irrigation. Huge reservoirs are built by damming major rivers, which can supply water to large irrigation areas measuring thousands of hectares. Small and medium sized reservoirs have a far more modest capacity, with enough water to irrigate 10 to 100 ha in one season. This training manual focuses on the latter type of reservoir.

Water Availability in Lakes and Reservoirs

Natural Lakes: What has already been explained for rivers, applies also to lakes. The amount of irrigation water required must be less than the quantity of water which is available in the lake.

When lakes are supplied with water from rivers, their level may vary from one year to another depending on the river. When lakes are supplied from ground-water, their water levels fluctuate more slowly, following the level of the ground-water.

The amount of lake water available for irrigation may be calculated as the difference between the water available in the lake and the amount lost due to evaporation and seepage. For large irrigation schemes, local or national authorities will be responsible for evaluating the available quantity of water. They will have data on river flows and ground-water supplies. They can also estimate the evaporation and percolation losses and will have information on other uses for the water.

Reservoirs: The amount of water that can be stored in a reservoir depends on:

- river discharge (which in turn depends on rainfall)
- height of the dam (when closing the river in a valley)
- area of the reservoir (in the case of an artificially excavated storage area).

A dam built in a valley will block the river or stream and water can then be stored behind it. The deeper and wider the valley, the more water that can be stored. This is called on-stream storage.

Water can also be stored off-stream by diverting the flow into a natural depression or dead river branch alongside the river, or by constructing a storage reservoir. This is similar to the lake situation.

Evaporation from open water can easily reach 7 mm per day in arid or semi-arid countries. This equals 5 cm per week and 20 cm per month. The amount of water lost by evaporation can be considerable, particularly in reservoirs which are large and shallow. Therefore, irrigation from shallow lakes and reservoirs should be started as soon as possible after the rainy season. Wait too long, and the reservoir may well have dried up.

Because the cost of building and maintaining a reservoir can be significant, expert help should be sought in planning and constructing a dam. Specialists can calculate the volume of runoff water, anticipate water losses from the reservoir and estimate the quantity of water available for irrigation. They should also be able to provide correct engineering designs and sound construction techniques to ensure the stability of the dam, including spillway arrangements to pass excess water safely downstream without harming the dam.

Water Tapping from a Reservoir

Water stored in a valley usually has a higher level than the valley bottom downstream of the dam. Because of this difference in level, the valley can be irrigated by a gravity system.

Water can be taken from the reservoir via a concrete or steel pipe. This pipe connects the reservoir to an irrigation canal downstream. A valve is usually located on the upstream end of the pipe to control the discharge water into the canal.

Pumping from a Lake or Reservoir: The fields located around the reservoir upstream of a dam or surrounding a natural lake are higher than the reservoir or lake's water table. Irrigation here will only be possible with the help of pumps. The water level in the reservoir will usually be highest at the end of the rainy season, and lowest at the end of the dry season or the irrigation season. Pumps installed at reservoirs and lakes must be able to handle these fluctuations, which are not only vertical, but which are even more pronounced horizontally, because the water recedes back to the lowest parts of the reservoir.

A dead branch of a river can also be made to function as a reservoir. The branch is filled with water during the wet season and closed off during the dry season so that the stored water may be used. Due to the low water level, pumps will normally be needed to irrigate fields from such a reservoir.

Ground-water

Ground-water is an important source of irrigation water, especially for small-scale irrigation projects. Because ground-water is only available below ground level, it must be lifted, or pumped before it can be used. Pumping ground-water from wells is a well known method of utilising ground-water the world over. To understand how ground-water functions, think of it as a series of lakes below the surface of the earth. The earth is built up of different layers - sand, gravel, clay, rock, etc. The layers of rock or compact clay cannot store water as they are solid, rather than porous. The layers of coarse sand and gravel, on the other hand, contain many pores and cracks, which allow rainfall to enter the soil and percolate from the surface. These porous layers filled with water are called aquifers. Ground-water flows, in most cases, slowly to the lower parts. Where the aquifer meets the surface, the ground-water flows out of the soil into, for instance.

Groundwater may be found close to the surface or at profound depths. In coastal plains the groundwater is often brackish or saline due to the proximity of the sea. Inland groundwater may also be brackish in places where the soil contains many soluble salts. Such water cannot be used for irrigation.

Groundwater Availability

The availability of groundwater is less irregular than that of small rivers. Aquifers at shallow depths, however, are likely to be very thin with a limited storage capacity for groundwater. One may start pumping the water during the rainy season, after the aquifer has been replenished, and find a dry well after a few months. Generally, local farmers know the wells in their areas. When digging a new well, it is always wise to consult local farmers and benefit from their experience of existing wells in the neighbourhood.

The total water availability from shallow groundwater is determined by the number of wells and the capacity of these wells, or by the capacity of the pumps installed, whichever is smaller.

When a borehole is driven to access deep groundwater, the engineer always carries out pump tests to measure the capacity of the well. The volume of water stored in deeper aquifers is quite large, so that the only limitation to water availability is the capacity of the well and the pump to be installed. Excessive exploitation of the groundwater by many users, however, will bring about a decline in the water table. To ensure a stable supply of groundwater, the rate of use should not exceed the rate of recharge.

When pumps and other equipment break down, the resulting lack of water can be critical for plant growth. Ideally, a spare pump should be kept on hand in case of breakdowns in the main pumping system, as it can take a long time to repair or replace equipment.

Pumping from Wells

To tap groundwater, a well must be dug deeper than the groundwater level. The groundwater then seeps through the pores of the surrounding soil or porous rock into the well until the level of water in the well is the same as the level of the groundwater.

Shallow Groundwater: When the groundwater lies within a few metres of the surface, exploitation is possible with shallow wells which are mostly dug by hand. These wells normally have a diameter of 1 metre or more and are dug below the groundwater table. Water is pumped from these wells, often using human or animal power but, increasingly, with small diesel-powered pumps. The amount of water that can be abstracted from shallow wells is limited, and, as a result, the areas which are irrigated from these water sources will also be limited.

Deep Groundwater: When the groundwater level is very deep, constructing a hand-dug well becomes impossible. Deep wells must be drilled into the ground. Generally, submersible pumps are installed below the groundwater table to lift water to the surface. They may be driven by an electric motor or a diesel-powered engine on the surface with a long vertical shaft, or by a submerged electric motor inside a waterproof casing.

In case A, both pump and motor are submerged, while in case B, only the pump is submerged. The case B pump is driven by an electric motor located at the surface. In both cases, pumped water is pushed up the riser pipe to the surface.

Scheme Irrigation Water Need and Supply

This chapter deals with the irrigation water need of an entire irrigation area, SIN_{gross}, and the supply of irrigation water to the entire irrigation area, SIN_{op}. The SIN_{gross} is the continuous flow of water required for good crop production during the irrigation season. SIN_{op} is the flow of water actually applied to the irrigation area, continuously or at intervals after taking the operational criteria into account. Because there is always some water stored in the soil (rootzone of the crop), or on the field (in case of paddy rice), SIN_{gross} can be supplied at intervals without affecting crop production. However, over the long run, the supplied volumes of water should equal the needed volumes of water.

The SIN_{gross} may be determined quickly, just to get a rough idea of the volume of water needed. This quick method, which is explained, is called the approximate method and is suitable for preliminary planning only, or when there is a lack of reliable data. The second, the calculation method is more accurate and thus suitable for the detailed planning of sound water management. The calculation method, uses fairly detailed information on the irrigation area's cropping system.

The Approximate Method

The approximate method of determining the SIN_{gross} assumes a constant average irrigation need, IN_{net}, for the entire season. The most simplified standard value of this IN_{net} is 1 litre per second per hectare. This is equivalent to a daily water requirement of 8.6 mm. When the daily water requirement is 4.3 mm, the irrigation need would be 0.5 l/s.

The approximate values of the irrigation need for an entire area, SIN_{net} may be calculated by multiplying the IN_{net} with the area, Area (in hectares). Therefore, the equation representing the approximate Net Scheme Irrigation Need is:

$$SIN_{net}\ (l/s) = Area\ (ha) \times IN_{net}\ (l/s/ha)$$

The net scheme irrigation need, SIN_{net} is the amount of water needed to meet crop water needs of an entire scheme minus the effective rainfall. Water lost during delivery must be added before the gross scheme irrigation need, SIN_{gross}, can be determined.

The SIN_{gross} is calculated by dividing SIN_{net} by the overall scheme irrigation efficiency (e), which is the product of the conveyance efficiency (e_c) and the field application efficiency (e_a). The subject of irrigation efficiencies is discussed at length in Annex 1 in Training Manual 4, "Irrigation Scheduling".

The present series of training manuals is mainly concerned with small irrigation schemes and tertiary units of larger schemes which use gravity irrigation methods. Gravity irrigation of sandy soils is not widespread. Therefore, a conveyance efficiency of about 85 per cent and an application efficiency of about 60 per cent have been selected, which result in an overall area irrigation efficiency of about 50 per cent. This value for the overall efficiency will be used in all further calculations in this manual. It should also be kept in mind that the values mentioned above are only indicative values.

Once the scheme irrigation efficiency (e, expressed in percentages) is known, SIN_{gross} can be calculated with the following formula:

$$SIN_{gross}\ (l/s) = 100/e \times SIN_{net}\ (l/s)$$

The formula above shows that as the efficiency drops (becomes smaller), more water has to be supplied to satisfy the irrigation need of a scheme.

Operational Criteria: The amount of water that is supplied to an irrigation area should be equal to the gross scheme irrigation water need.

This would mean that, if the same flow of water is supplied continuously, second after second, hour after hour, and day after day without interruption throughout the entire growing season, the SIN_{gross} which is expressed in a continuous flow of litres per second would have been completely supplied, and SIN_{op}, operational scheme irrigation need, would equal SIN_{gross}.

However, water is not always supplied continuously. If irrigation water is applied only 12 out of a possible 24 hours per day, then during these 12 hours twice as much water must be given to supply the same total. Further, if irrigation does not take place every day, more water must be supplied on irrigation days in order to maintain the same total quantity per week.

The ratios of hours per day and days per week during which the irrigation system functions are called, in this manual, the operational criteria. By using these operational criteria, the required scheme irrigation supply can be determined with the following formula:

$$SIN_{op}\ (l/s) = SIN_{gross}/T_{op}$$

$$T_{op} = d/7 \times h/24$$

where:

SIN_{op} = operational scheme irrigation need

T_{op} = operational criteria

d = number of irrigation days per week

h = number of irrigation hours per day

SIN_{gross} = gross scheme irrigation need

The formula above shows that the fewer the number of irrigation applications per week and the fewer the number of hours of irrigation per day, the larger the scheme's supply should be, and the larger should also be the capacity of the irrigation system.

Cropping Pattern

A much more accurate prediction of a scheme's irrigation flow requirements can be made, if it is based on the scheme's cropping pattern. The cropping pattern, or cropping schedule of an irrigation area provides information, for a period of at least one season, on three important elements:

- which crops are grown
- when are they cultivated
- how many hectares of each crop are grown.

This information is written down, in detail, let us say for the area of one farmer during one year, and from this record there emerges a pattern. For example, the cropping pattern for farm A:

- onions cultivated from April 15 to September 15 on 1 ha
- potatoes from October 15 to February 15 on 1/2 ha, on the same plot as the onions
- cotton from July 10 to January 20 on 1 and 1/2 ha.

Information on the type of crop and the period of cultivation can be visualised in a line diagram.

This crop calendar diagram shows that farmer A starts with onion cultivation in mid-April which he then harvests mid-September. Within one month he will plant potatoes on the same plot, which he expects to uproot in mid-February. Farmer A has a second plot on which he grows cotton over a six-month period from mid-July to the end of January of the following year.

If the area of each crop has to be shown, the lines representing individual crops should be replaced by bars where the height of each bar is a measure of the area of each crop. In the example below, for instance, an area of one hectare is represented by a bar-height of about 0.5 centimetre. The length of the bar indicates the period in which the crop is grown. A diagram with crop bars rather than crop lines is called a cropping pattern.

The diagram can indicate more crop husbandry dates, such as crop manuring and periods of weeding and transplanting, etc. It is also possible to draw up a cropping pattern diagram for larger areas, including a group of farmers, or an entire irrigation scheme.

The only difference is that with a single farm, it was supposed that the crop was sown or planted on one specific day. The same assumption cannot be made in the case of a larger irrigation scheme. In fact, there are several reasons why farmers do not all start planting on the same day on an irrigation scheme:

- labour and machinery needed to prepare the land is in short supply, and each has to wait his turn;
- a farmer is not ready due to other activities, so he postpones planting the crop for a while;
- also, it is impossible to supply all the fields on an irrigation scheme with the water required to start a crop on the same day. The larger the scheme's area, the longer the time it takes to serve all the fields.

The cultivation of any crop on an irrigation scheme is thus spread over a certain period of time. This is called staggered cultivation or a staggered cropping pattern.

When, for example, the first farmer starts with onion cultivation on 1 April, and other farmers follow more or less regularly on the days after until the last farmer has planted his onions at the end of April, the crop calendar for onions is staggered over one month. In a cropping pattern diagram this staggering is depicted as below. Each line in the diagram represents a farm, or a certain area, and the day it starts onion cultivation. Of course; when crop planting dates are staggered, so also are the end of season dates.

In a normal cropping pattern diagram the subsequent lines are simplified, forming a parallelogram bar. The lower side represents the first farmer, or the area which is planted earliest. The upper side represents the last area to be planted and cropped. As with the cropping pattern diagram for one farmer, the extent of the area cultivated for each type of crop, across the entire scheme, can be indicated with the height of the parallelogram serving as a measure. The height of the vertical axis of the diagram represents the total area of an irrigation scheme, for example a scheme of 150 ha. Onions are cultivated on 60 ha and they have a growing period of 5 months. The onion planting season starts in the beginning of April and is staggered over one month. Potatoes, with a growing period of 4 months, are grown on 30 ha starting 1 October. The cultivation period is staggered over one month. Cotton, with a growth period of 190 days, is grown on 75 ha from the beginning of July. Cotton cultivation is staggered over a period of 45 days.

An area which is empty at any given moment in the cropping pattern diagram is not cultivated at that time. For example, during the month of March nothing is being cultivated on the scheme; in mid-September 30 ha of onions have already been harvested and 15 ha of land have been left fallow.

The Calculation Method

The irrigation water need and the actual scheme irrigation flow requirement were estimated using the approximate method, which is based on a rule of thumb which says that one hectare of irrigated land needs a fixed flow of water during the irrigation season.

When accurate crop and climatic data are available, a much more precise prediction may be made of the gross scheme irrigation need, SIN_{gross}, while the operational scheme irrigation need, SIN_{op}, can be finely tuned to the real irrigation need. However, when a scheme has many crops, this more precise method also requires more complex calculations. A calculation method example is presented in this section to help field technicians and extension agents understand and become familiar with the underlying principles and the basic elements involved in calculating scheme water needs.

The calculation of an individual crop's irrigation water need, IN_{net}, has already been explained in Training Manual 3, "Irrigation Water Needs". Using this calculation as a point of departure, which ends with the net irrigation need of an individual crop expressed in millimetres of water depth per month (mm/mon), this section will focus on determining the net scheme irrigation need SIN_{net} (in litres per second) of an entire irrigation scheme with a variety of crops throughout the year.

Scheme Irrigation need for a Simple Cropping System

Suppose, as in the former section, that 60 ha of onions are planted on an irrigation scheme of 150 ha and that the planting of onions is spread out evenly over one month. Planting is begun on 1 April and finished on 30 April.

The irrigation water need of the earliest planted onions during the month of April is 98 mm, which is 3.3 mm/day for 30 days. Onions which are not planted until April 15 do not need water during the first 15 days of April. For the remaining 15 days of April, the irrigation water need is 15 x 3.3 mm, or 49 mm. Onions which are planted on the last day of April do not need any irrigation water in April at all. This example indicates that the irrigation water need of the onions which are planted in between the earliest and the latest planted onions, lies in between the irrigation water need of the earliest and latest planted onions. In general, the average irrigation water need of an area cropped with onions, or any other crop, is the average of the irrigation water need of the earliest (IN_e) and the irrigation water need of the latest (IN_l) planting. This is also expressed in the formula below:

$$IN_{av} = \frac{IN_e + IN_1}{2}(mm / month)$$

This monthly irrigation need, IN_{av}, is presented as a water layer which is applied during that month on a certain area. However, water flows are not actually expressed in millimetres of water covering a certain area over a certain period of time, but in litres per second, or cubic metres per second. Therefore the IN_{av} per month should be converted into a current irrigation water flow. SIN_{net} in litres per second (net scheme irrigation need) can be calculated as follows:

$$SIN_{net} = \frac{IN_{av}\left(mm / month\right) \times Area\left(ha\right) \times 10000(1 / mm.ha)}{30\ (days / month) \times 86400(s / day)} (l / s)$$

or

$$SIN_{net} = \frac{IN_{av}(mm / month) \times Area(ha)}{260} (l / s)$$

When the monthly irrigation need of an area with staggered crop cultivation must be determined, the irrigation needs of the earliest and of the latest plantings must be calculated.

Matching Water Needs and Supply

This chapter will explain why the irrigation need and the irrigation supply should always be compared, particularly when the important elements in the scheme are likely to change, or when water shortages occur, as is all too frequently the case. A number of methods are discussed which can help diminish the danger of water shortages. This part of the manual is intended as a guide to the practical problems of water shortage. Moreover, the information offered here should provide field technicians with a water management tool with which to anticipate and solve irrigation problems at the more complex level of the agricultural scheme.

Command Area

The command area, CA, of a water source is the extent of area which can be reliably irrigated from that source. Reliable irrigation means that the availability of water is always larger than or equal to the irrigation need of a scheme. If the command area of a water source is larger than the actual area to be irrigated, there should be no problem of water shortage. When the supply of a water source in a certain month is known, e.g. 250 litres per second, and the gross irrigation need per hectare for the same month is estimated, for instance at 1.8 litres per second per hectare, then the command area can be calculated. With a water supply of 250 l/s and a need of 1.8l/s per hectare, one can irrigate 250/1.8, or 139 ha. The formula used to calculate the command area is scheme water supply (SWS) divided by gross irrigation need (IN_{gross}) or

$$CA = \frac{SWS}{IN_{gras}} (ha)$$

When the supply of a water source is not constant over the months and seasons of the year, then the command area will also vary. Imagine, for example, a SWS from a creek with the monthly values indicated in the schedule below (these values will be used for all of the examples throughout this chapter).

Suppose that the farmers grow sugar cane on an area measuring 420 hectares. This crop has a growing period of 12 months. In this case, only during one month, February, would there be enough water available to irrigate that total area. In other words irrigation would be highly unreliable. If planting were confined to an area of not more than 140 ha, the sugar cane could be supplied with sufficient water throughout the year. Irrigation in this case is reliable.

Matching SIN_{gross}, and SWS

IN_{gross} was defined as the gross irrigation need per hectare expressed in litres per second per hectare. When we know the total irrigated area of a scheme, we can multiply the IN_{gross} with the area, and we find the gross scheme irrigation need, SIN_{gross}

SIN_{gross} (l/s) =IN_{gross} (l/s/ha) x Area (ha)

The SIN_{gross} is the continuous water flow that the irrigation scheme needs, the SWS is the water flow that can be delivered by the water source. In this section, five possible means or methods are discussed of matching the SIN_{gross} with the SWS, when the irrigation supply is less than the amount required.

Method 1: Enlarge the SWS if the SIN_{gross} is too large.

Enlarging the SWS is only possible when the water source is not fully exploited. This may be the result of a pumping system that does not have enough power to lift the required water flow, or an intake structure or supply canal that is too small or placed too high above the water level in the river or reservoir.

These situations can be improved and the supply increased. Of course, installing a new pump or intake will entail expenditures and other efforts. Enlarging the SWS is impossible when the river flow itself is too small, except unless a very expensive reservoir is constructed.

If irrigation water is being pumped from a well, it may be the case that the capacity of the well is limited. In this case, the well has to be deepened, or a second well and pump installed.

Method 2: Diminish the required SIN_{op} by changing the operational criteria.

When irrigation water is supplied intermittently, the water source is not used full-time. For instance, river water flows away during the hours or days that the river is not tapped. The exploitation of the river is not optimal in such a case. Moreover, when irrigation applications are prolonged or irrigation intervals are shortened, the required SIN_{op} will diminish.

If, however, the pump is already used most of the time, or farmers do not want to change the irrigation schedule, which are related to the operational criteria, another solution must be found.

Method 3: Matching by advancing or postponing the growing season.

In the case of a variable river flow, the SWS might be too low at the beginning or at the end of the growing season. If a water shortage occurs at the beginning of the season, one may consider postponing the growing season by a month or so. If water problems are more likely to occur at the end of the season, it may help to plant the crop earlier.

Method 4: Stagger the growing season.

Method 3 consisted of either advancing or postponing the growing period for the entire irrigation scheme. By doing so, it is sometimes possible to avoid the risk of water shortages at the beginning or at the end of the irrigation season. With Method 4, the growing season is shifted, but only for a part of scheme. In planning this operation carefully, the SIN_{gross} can be fine-tuned to an even greater extent to the SWS.

In the example priority was given to planting as large an area as possible as soon as possible. This is the best solution if farmers on the scheme want to start the rice season as early as possible. However, if they were to prefer planting the rice as late in the season as possible, and harvest in July or August, we will need to determine the critical command area for the last 4 months. Areas where there is not enough water available during the last four months, should be planted in advance.

Lake

A lake is a body of relatively still fresh or salt water of considerable size, localised in a basin, that is surrounded by land. Lakes are inland and not part of the ocean and therefore are distinct from lagoons, and are larger and deeper than ponds. Lakes can be contrasted with rivers or streams, which are usually flowing. However most lakes are fed and drained by rivers and streams.

Natural lakes are generally found in mountainous areas, rift zones, and areas with ongoing glaciation. Other lakes are found in endorheic basins or along the courses of mature rivers. In some parts of the world there are many lakes because of chaotic drainage patterns left over from the last Ice Age. All lakes are temporary over geologic time scales, as they will slowly fill in with sediments or spill out of the basin containing them.

Many lakes are artificial and are constructed for industrial or agricultural use, for hydro-electric power generation or domestic water supply, or for aesthetic or recreational purposes.

Etymology, Meaning, and Usage of "Lake"

The word *lake* comes from Middle English *lake* ("lake, pond, waterway"), from Old English *lacu* ("pond, pool, stream"), from Proto-Germanic **lakô* ("pond, ditch, slow moving stream"), from the Proto-Indo-European root **leg'-*

("to leak, drain"). Cognates include Dutch *laak* ("lake, pond, ditch"), Middle Low German *lâke* ("water pooled in a riverbed, puddle"), German *Lache* ("pool, puddle"), and Icelandic *lækur* ("slow flowing stream"). Also related are the English words *leak* and *leach*.

There is considerable uncertainty about defining the difference between lakes and ponds, and no current internationally accepted definition of either term across scientific disciplines or political boundaries exists. For example, limnologists have defined lakes as water bodies which are simply a larger version of a pond, which have wave action on the shoreline or where wind-induced turbulence plays a major role in mixing the water column. None of these definitions completely excludes ponds and all are difficult to measure. For this reason there has been increasing use made of simple size-based definitions to separate ponds and lakes. One definition of *lake* is a body of water of 2 hectares (5 acres) or more in area, however others have defined lakes as waterbodies of 5 hectares (12 acres) and above, or 8 hectares (20 acres) and above. Charles Elton, one of the founders of ecology, regarded lakes as waterbodies of 40 hectares (99 acres) or more. The term *lake* is also used to describe a feature such as Lake Eyre, which is a dry basin most of the time but may become filled under seasonal conditions of heavy rainfall. In common usage many lakes bear names ending with the word *pond*, and a lesser number of names ending with *lake* are in quasi-technical fact, ponds.

In lake ecology the environment of a lake is referred to as *lacustrine*. Large lakes are occasionally referred to as "inland seas," and small seas are occasionally referred to as lakes, such as Lake Maracaibo, which is actually a bay. Larger lakes often invert the word order, as in the names of each of the Great Lakes, in North America.

Only one lake in the English Lake District is actually called a lake; other than Bassenthwaite Lake, the others are all *meres* or *waters*. Only six bodies of water in Scotland are known as lakes (the others are lochs): the Lake of Menteith, the Lake of the Hirsel, Pressmennan Lake, Cally Lake near Gatehouse of Fleet, the saltwater Manxman's Lake at Kirkcudbright Bay and The Lake at Fochabers. Of these only the Lake of Menteith and Cally Lake are natural bodies of fresh water.

Distribution of Lakes

The majority of lakes on Earth are fresh water, and most lie in the Northern Hemisphere at higher latitudes. More than 60 percent of the world's lakes are in Canada; this is because of the deranged drainage system that dominates the country.

Finland is known as *The Land of the Thousand Lakes*, (actually there are 187,888 lakes in Finland, of which 60,000 are large), and the U.S. state of Minnesota is known as *The Land of Ten Thousand Lakes*. The license plates of the Canadian province of Manitoba used to claim *100,000 lakes* as one-upmanship on Minnesota, whose license plates boast of its *10,000 lakes.*

Most lakes have at least one natural outflow in the form of a river or stream, which maintain a lake's average level by allowing the drainage of excess water. Some lakes do not have a natural outflow and lose water solely by evaporation or underground seepage or both. They are termed endorheic lakes. Many lakes are artificial and are constructed for hydro-electric power generation, aesthetic purposes, recreational purposes, industrial use, agricultural use or domestic water supply.

Evidence of extraterrestrial lakes exists; "definitive evidence of lakes filled with methane" was announced by NASA as returned by the Cassini Probe observing the moon Titan, which orbits the planet Saturn.

Globally, lakes are greatly outnumbered by ponds: of an estimated 304-million standing water bodies worldwide, 91 percent are 1 hectare (2.5 acres) or less in area. Small lakes are also much more numerous than large lakes: in terms of area, one-third of the world's standing water is represented by lakes and ponds of 10 hectares (25 acres) or less. However, large lakes contribute disproportionately to the area of standing water with 122 large lakes of 1,000 square kilometres (390 sq mi, 100,000 ha, 247,000 acres) or more representing about 29 percent of the total global area of standing inland water.

Origin of Natural Lakes

There are a number of natural processes that can form lakes. A recent tectonic uplift of a mountain range can create bowl-shaped depressions that accumulate water and form lakes. The advance and retreat of glaciers can scrape depressions in the surface where water accumulates; such lakes are common in Scandinavia, Patagonia, Siberia and Canada. The most notable examples are probably the Great Lakes of North America.

Lakes can also form by means of landslides or by glacial blockages. An example of the latter occurred during the last ice age in the U.S. state of Washington, when a huge lake formed behind a glacial flow; when the ice retreated, the result was an immense flood that created the Dry Falls at Sun Lakes, Washington.

Salt lakes (also called saline lakes) can form where there is no natural outlet or where the water evaporates rapidly and the drainage surface of the water table has a higher-than-normal salt content. Examples of salt lakes include Great Salt Lake, the Aral Sea and the Dead Sea.

Small, crescent-shaped lakes called oxbow lakes can form in river valleys as a result of meandering. The slow-moving river forms a sinuous shape as the outer side of bends are eroded away more rapidly than the inner side. Eventually a horseshoe bend is formed and the river cuts through the narrow neck. This new passage then forms the main passage for the river and the ends of the bend become silted up, thus forming a bow-shaped lake. Crater lakes are formed in volcanic craters and calderas which fill up with precipitation more rapidly than they empty via evaporation. Sometimes the latter are called caldera lakes, although often no distinction is made. An example is Crater Lake in Oregon, located within the caldera of Mount Mazama. The caldera was created in a massive volcanic eruption that led to the subsidence of Mount Mazama around 4860 BC. Gloe Lakes are freshwater lakes that have emerged when the water they consists of has been separated, not considerably long before, from the sea as a consequence of post-glacial rebound. Some lakes, such as Lake Jackson in Florida, USA, come into existence as a result of sinkhole activity. Lake Vostok is a subglacial lake in Antarctica, possibly the largest in the world. The pressure from the ice atop it and its internal chemical composition mean that, if the lake were drilled into, a fissure could result that would spray somewhat like a geyser. Most lakes are geologically young and shrinking since the natural results of erosion will tend to wear away the sides and fill the basin. Exceptions are those such as Lake Baikal and Lake Tanganyika that lie along continental rift zones and are created by the crust's subsidence as two plates are pulled apart. These lakes are the oldest and deepest in the world. Lake Baikal, which is 25-30 million years old, is deepening at a faster rate than it is being filled by erosion and may be destined over millions of years to become attached to the global ocean. The Red Sea, for example, is thought to have originated as a rift valley lake.

Types of Lakes

- *Periglacial Lake:* Part of the lake's margin is formed by an ice sheet, ice cap or glacier, the ice having obstructed the natural drainage of the land.
- *Subglacial Lake:* A lake which is permanently covered by ice. They can occur under glaciers, ice caps or ice sheets. There are many such lakes, but Lake Vostok in Antarctica is by far the largest. They are kept liquid because the overlying ice acts as a thermal insulator retaining energy introduced to its underside by friction, by water percolating through crevasses, by the pressure from the mass of the ice sheet above or by geothermal heating below.
- *Glacial Lake:* a lake with origins in a melted glacier, such as a kettle lake.
- *Artificial Lake:* A lake created by flooding land behind a dam, called an impoundment or reservoir, by deliberate human excavation, or by the flooding of an excavation incident to a mineral-extraction operation such as an open pit mine or quarry. Some of the world's largest lakes are reservoirs like Hirakud Dam in India.

- *Endorheic Lake, Terminal or Closed:* A lake which has no significant outflow, either through rivers or underground diffusion. Any water within an endorheic basin leaves the system only through evaporation or seepage. These lakes, such as Lake Eyre in central Australia or the Aral Sea in central Asia, are most common in desert locations.
- *Meromictic Lake:* A lake which has layers of water which do not intermix. The deepest layer of water in such a lake does not contain any dissolved oxygen. The layers of sediment at the bottom of a meromictic lake remain relatively undisturbed because there are no living aerobic organisms.
- *Fjord Lake:* A lake in a glacially eroded valley that has been eroded below sea level.
- *Oxbow Lake:* A lake which is formed when a wide meander from a stream or a river is cut off to form a lake. They are called "oxbow" lakes due to the distinctive curved shape that results from this process.
- *Rift Lake or Sag Pond:* A lake which forms as a result of subsidence along a geological fault in the Earth's tectonic plates. Examples include the Rift Valley lakes of eastern Africa and Lake Baikal in Siberia.
- *Underground Lake:* A lake which is formed under the surface of the Earth's crust. Such a lake may be associated with caves, aquifers or springs.
- *Crater Lake:* A lake which forms in a volcanic caldera or crater after the volcano has been inactive for some time. Water in this type of lake may be fresh or highly acidic, and may contain various dissolved minerals. Some also have geothermal activity, especially if the volcano is merely dormant rather than extinct.
- *Lava Lake:* A pool of molten lava contained in a volcanic crater or other depression. Lava lakes that have partly or completely solidified are also referred to as lava lakes.
- *Former:* A lake which is no longer in existence. Such lakes include prehistoric lakes and lakes which have permanently dried up through evaporation or human intervention. Owens Lake in California, USA, is an example of a former lake. Former lakes are a common feature of the Basin and Range area of southwestern North America.
- *Ephemeral Lake:* A seasonal lake that exists as a body of water during only part of the year.
- *Intermittent Lake:* A lake with no water during a part of the year.

- *Shrunken:* Closely related to *former* lakes, a shrunken lake is one which has drastically decreased in size over geological time. Lake Agassiz, which once covered much of central North America, is a good example of a shrunken lake. Two notable remnants of this lake are Lake Winnipeg and Lake Winnipegosis.
- *Eolic Lake:* A lake which forms in a depression created by the activity of the winds.
- Vlei, in South Africa, shallow lakes which vary considerably with seasons.

Characteristics

Lakes have numerous features in addition to lake type, such as drainage basin (also known as catchment area), inflow and outflow, nutrient content, dissolved oxygen, pollutants, pH, and sedimentation. Changes in the level of a lake are controlled by the difference between the input and output compared to the total volume of the lake. Significant input sources are precipitation onto the lake, runoff carried by streams and channels from the lake's catchment area, groundwater channels and aquifers, and artificial sources from outside the catchment area. Output sources are evaporation from the lake, surface and groundwater flows, and any extraction of lake water by humans. As climate conditions and human water requirements vary, these will create fluctuations in the lake level. Lakes can be also categorised on the basis of their richness in nutrients, which typically affect plant growth. Nutrient-poor lakes are said to be *oligotrophic* and are generally clear, having a low concentration of plant life. *Mesotrophic lakes* have good clarity and an average level of nutrients. *Eutrophic* lakes are enriched with nutrients, resulting in good plant growth and possible algal blooms. *Hypertrophic* lakes are bodies of water that have been excessively enriched with nutrients. These lakes typically have poor clarity and are subject to devastating algal blooms. Lakes typically reach this condition due to human activities, such as heavy use of fertilizers in the lake catchment area. Such lakes are of little use to humans and have a poor ecosystem due to decreased dissolved oxygen.

Due to the unusual relationship between water's temperature and its density, lakes form layers called thermoclines, layers of drastically varying temperature relative to depth. Fresh water is most dense at about 4°C (39.2°F) at sea level. When the temperature of the water at the surface of a lake reaches the same temperature as deeper water, as it does during the cooler months in temperate climates, the water in the lake can mix, bringing oxygen-starved water up from the depths and bringing oxygen down to decomposing sediments. Deep temperate lakes can maintain a reservoir of cold water year-round, which allows some cities to tap that reservoir for deep lake water cooling.

Since the surface water of deep tropical lakes never reaches the temperature of maximum density, there is no process that makes the water mix. The deeper layer becomes oxygen starved and can become saturated with carbon dioxide, or other gases such as sulphur dioxide if there is even a trace of volcanic activity. Exceptional events, such as earthquakes or landslides, can cause mixing which rapidly brings the deep layers up to the surface and release a vast cloud of gas which lay trapped in solution in the colder water at the bottom of the lake. This is called a limnic eruption. An example is the disaster at Lake Nyos in Cameroon. The amount of gas that can be dissolved in water is directly related to pressure. As deep water surfaces, the pressure drops and a vast amount of gas comes out of solution. Under these circumstances carbon dioxide is hazardous because it is heavier than air and displaces it, so it may flow down a river valley to human settlements and cause mass asphyxiation.

The material at the bottom of a lake, or *lake bed*, may be composed of a wide variety of inorganics, such as silt or sand, and organic material, such as decaying plant or animal matter. The composition of the lake bed has a significant impact on the flora and fauna found within the lake's environs by contributing to the amounts and the types of nutrients available. A paired (black and white) layer of the varved lake sediments correspond to a year. During winter, when organisms die, carbon is deposited down, resulting to a black layer. At the same year, during summer, only few organic materials are deposited, resulting to a white layer at the lake bed. These are commonly used to track past paleontological events.

Limnology

Limnology is the study of inland bodies of water and related ecosystems. Limnology divides lakes into three zones: the *littoral zone*, a sloped area close to land; the *photic* or *open-water zone*, where sunlight is abundant; and the deep-water *profundal* or *benthic zone*, where little sunlight can reach. The depth to which light can reach in lakes depends on turbidity, determined by the density and size of suspended particles. A particle is in suspension if its weight is less than the random turbidity forces acting upon it. These particles can be sedimentary or biological in origin and are responsible for the colour of the water. Decaying plant matter, for instance, may be responsible for a yellow or brown colour, while algae may cause greenish water. In very shallow water bodies, iron oxides make water reddish brown. Biological particles include algae and detritus. Bottom-dwelling detritivorous fish can be responsible for turbid waters, because they stir the mud in search of food. Piscivorous fish contribute to turbidity by eating plant-eating (planktonivorous) fish, thus increasing the amount of algae. The light depth or transparency is measured by using a *Secchi disk*, a 20-cm (8 in) disk with alternating white and black quadrants. The depth at which the disk is no longer visible is the *Secchi depth*, a measure of transparency. The Secchi disk is commonly used to test for eutrophication. For a detailed look at these processes. A lake moderates the surrounding region's temperature and climate because water has a very high specific heat capacity (4,186 $J \cdot kg^{-1} \cdot K^{-1}$). In the daytime a lake can cool the land beside it with local winds, resulting in a sea breeze; in the night it can warm it with a land breeze.

How Lakes Disappear

The lake may be infield with deposited sediment and gradually become a wetland such as a swamp or marsh. Large water plants, typically totos, accelerate this closing process significantly because they partially decompose to form peat soils that fill the shallows. Conversely, peat soils in a marsh can naturally burn and reverse this process to recreate a shallow lake. Turbid lakes and lakes with many plant-eating fish tend to disappear more slowly. A "disappearing" lake (barely noticeable on a human timescale) typically has extensive plant mats at the water's edge. These become a new habitat for other plants, like peat moss when conditions are right, and animals, many of which are very rare. Gradually the lake closes and young peat may form, forming a fen. In lowland river valleys where a river can meander, the presence of peat is explained by the infilling of historical oxbow lakes. In the very last stages of succession, trees can grow in, eventually turning the wetland into a forest.

Some lakes can disappear seasonally. These are called intermittent lakes and can be found in karstic terrain. A prime example of an intermittent lake is Lake Cerknica in Slovenia or Lag Prau Pulte in Graubunden. Other intermittent lakes are only the result of above-average precipitation in a closed, or endorheic basin, usually filling dry lake beds. This can occur in some of the driest places on earth, like Death Valley. This occurred in the spring of 2005, after unusually heavy rains. The lake did not last into the summer, and was quickly evaporated. A more commonly filled lake of this type is Sevier Lake of west-central Utah. Sometimes a lake will disappear quickly. On 3 June 2005, in Nizhny Novgorod Oblast, Russia, a lake called Lake Beloye vanished in a matter of minutes. News sources reported that government officials theorised that this strange phenomenon may have been caused by a shift in the soil underneath the lake that allowed its water to drain through channels leading to the Oka River. The presence of ground permafrost is important to the persistence of some lakes. According to research published in the journal *Science* ("Disappearing Arctic Lakes", June 2005), thawing permafrost may explain the shrinking or disappearance of hundreds of large Arctic lakes across western Siberia. The idea here is that rising air and soil temperatures thaw permafrost, allowing the lakes to drain away into the ground.

Some lakes disappear because of human development factors. The shrinking Aral Sea is described as being "murdered" by the diversion for irrigation of the rivers feeding it.

Extraterrestrial Lakes

At present the surface of the planet Mars is too cold and has too little atmospheric pressure to permit the pooling of liquid water on the surface. Geologic evidence appears to confirm, however, that ancient lakes once formed on the surface.

It is also possible that volcanic activity on Mars will occasionally melt subsurface ice, creating large lakes. Under current conditions this water would quickly freeze and sublimate unless insulated in some manner, such as by a coating of volcanic ash. Only one world other than Earth is known to harbour lakes, Saturn's largest moon, Titan. Photographs and spectroscopic analysis by the Cassini–Huygens spacecraft show liquid ethane on the surface, which is thought to be mixed with liquid methane.

Jupiter's small moon Io is volcanically active due to tidal stresses, and as a result sulphur deposits have accumulated on the surface. Some photographs taken during the Galileo mission appear to show lakes of liquid sulphur on the surface. There are dark basaltic plains on the Moon, similar to lunar maria but smaller, that are called *lacus* (singular *lacus*, Latin for "lake") because they were thought by early astronomers to be lakes of water.

Notable Lakes

- Lake Michigan-Huron is the largest lake by surface area: 117,350 km^2. It also has the longest lake coastline in the world: 8,790 km. If Huron and Michigan are considered two lakes, Lake Superior is the largest lake, with 82,414 km^2. However, Huron still has the longest coastline at 6,157 km (2980 km excluding the coastlines of its many inner islands). The world's smallest geological ocean, the Caspian Sea, at 394,299 km^2 has a surface area greater than the six largest freshwater lakes combined, and it's frequently cited as the world's largest lake.
- The deepest lake is Lake Baikal in Siberia, with a bottom at 1,637 m. Its mean depth is also the greatest in the world (749 m).

It is also the world's largest lake by volume (23,600 km^3, though smaller than the Caspian Sea at 78,200 km^3), and the second longest (about 630 km from tip to tip).

- The longest lake is Lake Tanganyika, with a length of about 660 km (measured along the lake's centre line).

It is also the second largest by volume and second deepest (1,470 m) in the world, after lake Baikal.

- The world's oldest lake is Lake Baikal, followed by Lake Tanganyika (Tanzania).

- The world's highest lake is the crater lake of Ojos del Salado, at 6,390 metres (20,965 ft). The Lhagba Pool in Tibet at 6,368 m (20,892 ft) comes second.
- The highest large freshwater lake in the world is Lake Manasarovar in Tibet Autonomous Region of China.
- The world's highest commercially navigable lake is Lake Titicaca in Peru and Bolivia at 3,812 m (12,507 ft). It is also the largest freshwater (and second largest overall) lake in South America.
- The world's lowest lake is the Dead Sea, bordering Israel and Jordan at 418 m (1,371 ft) below sea level. It is also one of the lakes with highest salt concentration.
- Lake Huron has the longest lake coastline in the world: about 2980 km, excluding the coastline of its many inner islands.
- The largest island in a freshwater lake is Manitoulin Island in Lake Huron, with a surface area of 2,766 km^2. Lake Manitou, located on Manitoulin Island, is the largest lake on an island in a freshwater lake.
- The largest lake located on an island is Nettilling Lake on Baffin Island, with an area of 5,542 km^2 and a maximum length of 123 km.
- The largest lake in the world that drains naturally in two directions is Wollaston Lake.
- Lake Toba on the island of Sumatra is located in what is probably the largest resurgent caldera on Earth.
- The largest lake located completely within the boundaries of a single city is Lake Wanapitei in the city of Sudbury, Ontario, Canada. Before the current city boundaries came into effect in 2001, this status was held by Lake Ramsey, also in Sudbury.
- Lake Enriquillo in Dominican Republic is the only saltwater lake in the world inhabited by crocodiles.
- Lake Bernard, Ontario, Canada, is the largest freshwater lake in the world with no islands.
- The largest lake in one country is Lake Michigan, in the U.S.A. However, it is sometimes considered part of Lake Michigan-Huron, making the record go to Great Bear Lake, Northwest Territories, in Canada, the largest lake within one jurisdiction.
- The largest lake on an island in a lake on an island is Crater Lake on Vulcano Island in Lake Taal on the island of Luzon, The Philippines.

- The northernmost named lake on Earth is Upper Dumbell Lake in the Qikiqtaaluk Region of Nunavut, Canada at a latitude of 82°28'N. It is located only 5.2 kilometers southwest of Alert, the northernmost settlement in the world. There are also several small lakes located north of Upper Dumbell Lake, but they are all unnamed and only appear on very detailed maps.

Largest by Continent

The largest lakes (surface area) by continent are:

- Australia – Lake Eyre (salt lake)
- Africa – Lake Victoria, also the third-largest freshwater lake on Earth. It is one of the Great Lakes of Africa.
- Antarctica – Lake Vostok (subglacial)
- Asia – Lake Baikal (if the Caspian Sea is considered a lake, it is the largest in Eurasia, but is divided between the two geographic continents)
- Oceania – Lake Eyre when filled; the largest permanent (and freshwater) lake in Oceania is Lake Taupo.
- Europe – Lake Ladoga, followed by Lake Onega, both located in northwestern Russia.
- North America – Lake Michigan-Huron, which is hydrologically a single lake. However, lakes Huron and Michigan are often considered separate lakes, in which case Lake Superior would be the largest.
- South America – Lake Titicaca, which is also the highest navigable body of water on Earth at 3,821 m above sea level. The much larger Lake Maracaibo is considered by some to be the second-oldest lake on Earth, but since it lies at sea level and nowadays is a contiguous body of water with the sea, others consider that it has turned into a bay.

Wetland

A wetland is an area of land whose soil is saturated with water either permanently or seasonally. Wetlands are categorised by their characteristic vegetation, which is adapted to these unique soil conditions. The water found in wetlands can be saltwater, freshwater, or brackish. Wetlands include swamps, marshes, and bogs, among others.

Wetlands are found on every continent except Antarctica. One of the world's largest natural wetlands is the Pantanal, which straddles Brazil, Bolivia and Paraguay in South America. The main functions of wetlands are as water purification systems, flood control, shoreline stability, and as reservoirs of biodiversity. Wetlands may be *converted* for agriculture or development or *constructed* as a water management tool as in the recent developing field of Water Sensitive Urban Design.

The study of wetlands has recently been termed paludology in some publications. Wetlands also are known internationally as inland water, or coastal/marine ecosystems.

The UN Millennium Ecosystem Assessment determined that environmental degradation is more prominent within wetland systems than any other ecosystem on Earth. International conservation efforts and the development of rapid assessment tools are being used in conjunction with each other to inform people about wetland issues.

Technical Definitions

Wetlands have been categorised both as biomes and ecosystems. A patch of land that develops pools of water after a rain storm would not be considered a "wetland" though the land is wet.

Wetlands have unique characteristics: they are generally distinguished from other water bodies or landforms based on their water level and on the types of plants that thrive within them. Specifically, wetlands are characterised as having a water table that stands at or near the land surface either permanently or seasonally for a long enough period each year to support aquatic plants. A concise definition is a community composed of hydric soil and hydrophytes.

Wetlands have also been described as ecotones, providing a transition between dry land and water bodies. Mitsch and Gosselink write that wetlands exist "...at the interface between truly terrestrial ecosystems and aquatic systems, making them inherently different from each other, yet highly dependent on both."

In environmental-decision making, there are subsets of definitions that are agreed upon to make *regulatory* and *policy* decisions.

Ramsar Convention Definition

Under the Ramsar international wetland conservation treaty, an international agreement signed by 160 countries, wetlands are defined as follows:

- *Article 1.1:* "...wetlands are areas of marsh, fen, peatland or water, whether natural or artificial, permanent or temporary, with water that is static or flowing, fresh, brackish or salt, including areas of marine water the depth of which at low tide does not exceed six metres."
- *Article 2.1:* "[Wetlands] may incorporate riparian and coastal zones adjacent to the wetlands, and islands or bodies of marine water deeper than six metres at low tide lying within the wetlands".

Regional Definitions

In the United States, wetlands are defined as "those areas that are inundated or saturated by surface or groundwater at a frequency and duration sufficient to support, and that under normal circumstances do support, a prevalence of vegetation typically adapted for life in saturated soil conditions. Wetlands generally include swamps, marshes, bogs and similar areas". This definition has been used in the enforcement of the Clean Water Act. Some US states, such as Massachusetts and New York, have separate definitions that may differ from the federal government.

Characteristics

Wetlands vary widely due to local and regional differences in topography, hydrology, vegetation, and other factors, including human disturbance. Wetlands can be divided into two main classes: tidal and non-tidal areas.

Hydrology

Wetland hydrology is associated with the spatial and temporal dispersion, flow and physiochemical attributes of surface and ground water in its reservoirs. Based on hydrology, wetlands can be categorised as riverine, lacustrine and palustrine. Sources of hydrological flows into wetlands are predominately precipitation, surface water and ground water. Water flows out of wetlands by evapotranspiration, surface run-off and sub-surface water outflow. Hydrodynamics (the movement of water through and from a wetland) affects hydroperiods (temporal fluctuations in water levels) by controlling the water balance and water storage within a wetland.

Landscape characteristics control wetland hydrology and hydro-chemistry. The O_2 and CO_2 concentrations of water depend on temperature and atmospheric pressure. Hydro-chemistry within wetlands is determined by the pH, saline content, nutrients, conductivity, soil composition, hardness and the sources of water. Water chemistry of wetlands varies across landscapes and climatic regions. Wetlands are generally minerotrophic with the exception of bogs. Bogs receive their water from the atmosphere and therefore their water has low mineral ionic composition because ground water has a higher concentration of dissolved nutrients and minerals in comparison to precipitation.

The water chemistry of fens ranges from low pH and low minerals to alkaline with high accumulation of calcium and magnesium because they acquire their water from precipitation as well as ground water.

The Role of Salinity

Salinity has a strong influence on wetland water chemistry. In non-riverine wetlands, natural salinity is regulated by interactions between ground and surface water, which may be influenced by human activity.

Soil

Carbon is the major nutrient cycled within wetlands. Most nutrients, such as sulphur, phosphorus, carbon and nitrogen are found within the soil of wetlands. Anaerobic and aerobic respiration in the soil influences the nutrient cycling of carbon, hydrogen, oxygen and nitrogen, and the solubility of phosphorus thus contributing to the chemical variations in its water. Wetlands with low pH and saline conductivity may reflect the presence of acid sulphates and wetlands with average salinity levels can be heavily influenced by calcium or magnesium. Biogeochemical processes in wetlands are determined by soils with low redox potential.

Biota

The biota of a wetland system includes its vegetation zones and structure as well as animal populations and distribution which are highly dependent on water chemistry. The chemistry of water flowing into wetlands depends on the source of water and the geological material in which it flows through as well as the nutrients discharged from organic matter in the soils and plants at higher elevations in slope wetlands. Biota may vary within a wetland due to season or migratory animals.

Flora

There are four main groups of hydrophytes that found in wetland systems throughout the world.

Submerged water plants. This type of vegetation in found completely underwater. It may range from sea grasses in inter-tidal zones to ribbonweed, the most common native plant found within Australian waterways. Submerged wetland vegetation can grow in saline and fresh-water conditions and may still possess a peak growing season. Endemic submerged species provide a food source for native fauna, habitat for invertrebrates, and also possess filtration capabilities.

Floating water plants. Floating vegetation is usually small although it may take up a large surface area in a wetland system. These hydrophytes have small roots and are only found in slow-moving water with rich-nutrient level water Floating aquatic plants are a food resource for avian species. A common example seen in wetland systems is the Lily pad or the Australian native duckweed.

Emergent water plants. Emergent water plants can seen above the surface of the water but whose roots are completely submerged. These hydrophytes will store nutrients and will proliferate to counter-balance harmful algal blooms and invasive species. Common rush is a common native emergent water plan found in the continent of Australia. It can grow up to one metre high and flowers in spring and summer.

Surrounding trees and shrubs. Surrounding wetland vegetation can be riparian or forested. Riparian vegetation grows close to river bank or coastal area stabilising the shore, providing shade to regulate water temperature, and provides habitat to native communities. Trees surround many wetland areas promoting a carbon sink, stabilisation, wind breaks, and erosion control. The Paperbark tree in Australia is commonly found in still or slow-moving water in swamps and creeks.

Fauna

Fish: Fish are the more dependent on wetland ecosystems than any other type of habitat. Every single species of fresh and saltwater fish spend part of its life-cycle in a coastal wetlands such as estuaries and mangroves. 75 per cent of the United States' commercial fish and shellfish stocks depend solely on estuaries to survive. In addition through the wetlands of the coral reef system, critical hatchery and nursery grounds are maintained for many tropical fish species.

Amphibians: Frogs are the most crucial amphibian species in wetland systems. Frogs need both terrestrial and aquatic habitats in which to reproduce and feed. While tadpoles control algal populations, adult frogs forage on insects. Frogs are used as an indicator of ecosystem health due to their thin skin which absorbs both nutrient and toxins from the surrounding environment resulting in an above extinction rate in unfavourable and polluted environmental conditions.

Reptiles: Alligators and crocodiles are two common reptilian species. Alligators are found in fresh water along with the fresh water species of the crocodile. The saltwater crocodiles is found in estuaries and mangroves and can be seen in the coastline bordering the Great Barrier Reef in Australia. The Florida Everglades is the only place in the world where both crocodiles and alligators co-exist. Snakes, lizards, goannas, and turtles also can be seen throughout wetlands.

Mammals: Multiple small mammals as well as large herbivore and apex species such as the Florida Panther live within and around wetlands. The wetland ecosystem attracts mammals due to its prominent seed sources, invertebrate populations and numbers of small reptiles and amphibians.

Marsupials: The platypus (*Ornithorhynchus anatinus*) The platypus is found in eastern Australia living in freshwater rivers or lakes, and create burrows for shelter and protection. They are active mainly at night-time hours, and use their webbed feet for swimming. Platypuses feed on insect larvae, worms or other freshwater insects. They do so mainly at night, by the use of their bill. They turn up mud on the bottom of the lake or river, and with the help of their electroreceptors located on the bill, find many insects and freshwater insects. They store their findings in special pouches behind their bill, and are consumed upon returning to the surface. Insects and invertebrates: these species total more than half of the 100,000 known animal species. Insects and invertebrates can be submerged in the water or soil, on the surface and in the atmosphere.

Algae

Algae are diverse water plants that can vary in size, colour and shape. Algae occur naturally in habitats such as inland lakes, inter-tidal zones and damp soil and provide a dedicated food source for animals, fish, and invertebrates. There are three main groups of algae:

Plankton are algae which are microscopic, free-floating algae that make the water look green or brown. This algae is so tiny that on average, if fifty of these microscopic algae were lined up end-to-end, it would only measure one millimetre. The plankton algae are at the start of the food chain and are therefore an important food source for all other organisms. Filamentous algae which are long strands of algae cells that form floating mats. Chara and Nitella algae are upright algae that look like a submerged plant with roots.

Biota Importance

Many of the flora and fauna that are part of wetlands are disappearing. As most species are endemic to wetlands, biologists and other scientists routinely census wetland biota. The IUCN's red list can be accessed to determine if any wetland species existence is presently threatened.

Climates

Temperature

Temperatures vary greatly depending on the location of the wetland. Many of the world's wetlands are in temperate zones (midway between the North or South Pole and the equator). In these zones, summers are warm and winters are cold, but temperatures are not extreme. However, wetlands found in the tropic zone, around the equator, are warm year round. Wetlands on the Arabian Peninsula, for example, can reach 50°C (122°F) and would therefore be subject to rapid evaporation. In northeastern Siberia, which has a polar climate, wetland temperatures can be as low as –50°C (–58°F). In a moderate zone, such as the Gulf of Mexico, a typical temperature might be 11°C (51°F).

Rainfall

The amount of rainfall a wetland receives varies widely according to its location. Wetlands in Wales, Scotland, and Western Ireland typically receive about 1500 mm (or 60 in) per year. In some places in Southeast Asia, where heavy rains occur, they can receive up to 10,000 mm (about 200 in). In the northern areas of North America, wetlands exist where as little as 180 mm (7 inches) of rain fall each year.

Temporal variation:

- Perennial systems
- Seasonal systems
- Episodic (periodic or intermittent) systems
- Surface flow may occur in some segments, with subsurface flow in other segments
- Ephemeral (short-lived) systems
- Migratory species.

Human-disturbance:

- Encroachment
  - Drainage
  - Development
  - Over-grasing
  - Mining
  - Unsustainable water use
- Ecosystem Stress
  - Water Scarcity
  - Endangered species
  - Disruption of breeding grounds
  - Imbalance in sediment load and nutrient filtration.

The Peatswamps of South-east Asia

In Southeast Asia, peatswamp forests and soils are being drained, burnt, mined, and overgrazed contributing severely to climate change. As a result of peat drainage, the organic carbon that was built up over thousands of years and is normally under water, is suddenly exposed to the air. It decomposes and turns into carbon dioxide (CO_2), which is released into the atmosphere. Peat fires cause the same process and in addition create enormous clouds of smoke that cross international borders, such as happen every year in Southeast Asia.

Peatlands form only 3 per cent of all the world's land area, however, their degradation equals 7 per cent of all fossil fuel CO_2 emissions. Through the building of dams, Wetlands International is halting the drainage of peatlands in Southeast Asia, hoping to mitigate CO_2 emissions. Concurrent wetland restoration techniques include reforestation with native tree species as well as setting up community fire brigades. This sustainable approach can be seen in Central Kalimantan and Sumatra, Indonesia.

Aquaculture

Concerns are developing over farm fishing, a practice especially popular within Asia and the South Pacific and its influence on many small island developing states.

Ecosystem Services

The function of natural wetlands can be classified by their ecosystem benefits. The United Nations Millennium Ecosystem Assessment and Ramsar Convention found wetlands to be of biosphere significance and societal importance in the following areas:

- Flood control
- Groundwater replenishment
- Shoreline stabilisation and storm protection
- Water purification
- Reservoirs of biodiversity
- Wetland products
- Cultural values
- Recreation and tourism
- Climate change mitigation and adaptation.

> *"The economic worth of the ecosystem services provided to society by intact, naturally functioning wetlands is frequently much greater than the perceived benefits of converting them to 'more valuable' intensive land use – particularly as the profits from unsustainable use often go to relatively few individuals or corporations, rather than being shared by society as a whole".*
>
> – *Ramsar convention*

Unless otherwise cited, Ecosystem services is based on the following series of references.

Flood Control

Storage Reservoirs and Flood Protection. The wetland system of floodplains is formed from major rivers downstream from their headwaters. Notable river systems that produce large spans of floodplain include the Nile River (Africa), Mississippi River (USA), Amazon River (South America), Yangtze River (China), Danube River (Central Europe) and Murray-Darling River (Australia).

"The floodplains of major rivers act as natural storage reservoirs, enabling excess water to spread out over a wide area, which reduces its depth and speed. Wetlands close to the headwaters of streams and rivers can slow down rainwater runoff and spring snowmelt so that it doesn't run straight off the land into water courses. This can help prevent sudden, damaging floods downstream." Human-Impact. Converting wetlands through drainage and development have contributed to the issue of irregular flood control through forced adaption of water channels to narrower corridors due to loss of wetland area. These new channels must manage the same amount of precipitation causing flood peaks to be [higher or deeper] and floodwaters to travel faster.

Water management engineering developments in the past century have degraded these wetlands through the construction on artificial embankments. These constructions may be classified as dykes, bunds, levees, weirs, barrages and dams but serve the single purpose of concentrating water into a select source or area. Wetland water sources that were once spread over slowly over a large, shallow area are pooled into deep, concentrated locations. Loss of wetland floodplains results in more severe and damaging flooding. Catastrophic human impact in the Mississippi River floodplains was seen in death of several hundred individuals during a levee breach in New Orleans caused by Hurricane Katrina. Ecological catastrophic events from human-made embankments have been noticed along the Yangtze River floodplains after the where the middle of the river has become prone to more frequent and damaging flooding including the loss of riparian vegetation, a 30 per cent loss of the vegetation cover throughout the river's basin, a doubling of the percentage of the land affected by soil erosion, and a reduction in reservoir capacity through siltation build-up in floodplain lakes.

Highland wetlands. This photo of Pakistan demonstrates the interaction of wetlands within a larger ecosystem. A peat-lined lake lies adjacent to Karumbar Lake at 4,150m (13,615 ft). This makes highland wetlands such as these important to not only the wildlife that lives within the area but to all the lakes and waterways lying at a lower altitude and the communities living near them.

Groundwater Replenishment

The surface water seen within wetland systems represents a portion of the water cycle, which also includes atmospheric water and groundwater. Wetland systems' direct connection to groundwater regulates both the quantity and quality of groundwater. Wetland systems that have on permeable sediments or occur in areas with highly variable and fluctuating water tables have a role in groundwater replenishment or water recharge. Sediments that are porous allow water to filter down through the soil and overlying rock into aquifers, source of 95 per cent of the world's drinking water. Wetlands can also act as recharge areas when the surrounding water table is low and discharge zone when it is too high. Karst systems are a unique example of this system housing underground rivers influenced by precipitation causing the water table to fluctuate by 130 metres. Human-Impact. Groundwater is an important source of water for drinking and irrigation of crops. Over 1 billion people in Asia and 65 per cent of the public water sources in Europe from groundwater. Irrigation is a massive use of groundwater. 80 per cent of the world's groundwater is used for agricultural production in the United States, China, India, Iran, Pakistan and Bangladesh.

Unsustainable abstraction of groundwater is also a major concern. In the Commonwealth of Australia, water licensing and blank, blank is being implemented to control use of the water in the Murray-Darling floodplain. At a global scale, groundwater deficits and water scarcity is one of the most pressing concerns facing the 21st century.

Shoreline Stabilisation and Storm Protection

Tidal and inter-tidal wetland systems protect and stabilise coastal zones. Coral reefs provide a protective barrier to coastal shoreline. Mangroves stabilise the coastal zone from the interior and will migrate with the shoreline to remain adjacent to the boundary of the water. The main conservation benefit these systems have against storms and tidal waves is the ability to reduce the speed and height of waves and floodwaters.

Human-Impact: An estimated 200 million people live in low-lying coastal regions with estimates in the billions for communities that live along the coast or projected to live in urban coastal centres in the next 50 years .

The United Kingdom has begun the concept of managed coastal realignment. This management technique provides shoreline protection through restoration of natural wetlands rather than through applied engineering.

Water Purification

Nutrient Retention: Wetlands cycle both sediments and nutrients balancing terrestrial and aquatic ecosystems. A natural function of wetland vegetation is the take up and storage of nutrients found in the surrounding soil and water. These nutrients are retained in the system until the plant dies or is harvested by animals or humans. Wetland vegetation productivity is linked to the climate, wetland type, and nutrient availability. The grasses of fertile floodplains such as the Nile produce the highest yield including plants such as *Arundo donax*(giant reed), *Cyperus papyrus* (papyrus), *Phragmites* (reed) and *Typha* (cattail, bulrush).

Sediment Traps: Rainfall run-off is responsible for moving sediment through waterways. These sediments move towards larger and more sizable waterways through a natural process that moves water towards oceans. All types of sediments composed of clay, sand, silt, and rock may be carried into wetland systems through this process. Reedbeds or floodplain forests act as physical barriers to slow waterflow and trap sediment.

Water Purification: Many wetland systems possess biofilters, hydrophytes and organisms that in addition to nutrient up-take abilities have the capacity to remove toxic substances that have come from pesticides, industrial discharges and mining activities through filtration. The uptake by these plants stems, roots, and leaves has been shown to be many factors higher than the toxicity found in the wastewater. In particular, floating plants can absorb heavy metals. *Eichhornia crassipes* (water hyacinth), *Lemna* (duckweed) and *Azolla* (water fern) store iron and copper commonly found in wastewater. Many fast-growing plants rooted in the soils of wetlands such as *Typha* (cattail) and *Phragmites* (reed) also aid in the role of heavy metal up-take. Animals such as the oyster can filter more than 200 litres of water per day while grasing for food, removing nutrients, suspended sediments and chemical contaminants in the process.

Capacity: The ability of wetland systems to store nutrients and trap sediment is highly efficient and effective but each system has a threshold. An overabundance of nutrient input from fertilizer run-off sewage effluent or non-point pollution will cause eutrophication. Upstream erosion from deforestation can overwhelm wetlands making them shrink in size and see dramatic biodiversity loss through excessive sedimentation load. The capacity of wetland vegetation to store heavy metals is affected by waterflow, number of hectares (acres), climate, and type of plant.

Human-Impact. Introduced hydrophytes in different wetland systems can have devastating results. The introduction of *Eichhornia crassipes*, a native of South America into Lake Victoria in East Africa and *Lemna* into new areas in Queensland, Australia have overtaken critical wetlands suffocating the ecosystem due to its combination of phenomenal growth rate and floating nature.

Constructed Wetlands

Most natural wetlands systems are not designed for wastewater treatment (Pries 1994), however their potential in treating pollutants has been recognised by environmental engineers in constructing artificial wetlands. These treatment systems are highly-controlled environments that intend to mimic the occurrences of soil, flora and microorganism in natural wetlands to aid in treating wastewater effluent. Artificial wetlands have the ability to experiment with flow regimes, micro biotic composition and flora in order to produce the most efficient treatment process. Other advantages are the control of retention times and hydraulic channels. The most important factors of constructed wetlands are water flow processes and plant growth. Constructed wetland systems can be surface flow systems with free-floating macrophytes, floating-leaved macrophytes or submerged macrophytes; however typical free water surface systems are constructed with emergent macrophytes Constructed wetlands can be adapted to treat raw sewage, secondary domestic sludge, enhancing water quality of oxidation ponds' discharge, storm waters and mining, industrial and agricultural waste effluents as seen in wastewater management programmes in Kolkata (Calcutta), India and Arcata, California, USA.

'Kolkata's constructed wetland.' Kolkata is an example of how constructed wetlands are being utilised in developing countries.

Using the purification capacity of wetlands, the Indian city of Kolkata (Calcutta) has pioneered a system of sewage disposal that is both efficient and environmentally-friendly. Built to house one million people, Kolkata is now home to over 10 million, many living in slums. But the 8,000-hectare East Kolkata Wetlands Ramsar Site, a patchwork of tree-fringed canals, vegetable plots, rice paddies and fish ponds – and the 20,000 people that work in them – daily transform one-third of the city's sewage and most of its domestic refuse into a rich harvest of fish and fresh vegetables. For example, the Mudialy Fishermen's Cooperative Society is a collective of 300 families that lease 70 hectares into which wastewater from the city is released. Through a series of natural treatment processes – including the use of Eichhornia crassipes and other plants for absorbing oil, grease and heavy metals – the Cooperative has turned the area into a thriving fish farm and nature park. In 2005/06, the Cooperative sold fish worth over US$135,000 and shared income of more than US$55,000 among its members.

Reservoirs of Biodiversity

Wetland systems are extremely rich in biodiversity becoming a focal point at the Convention of Biodiversity and within the World Wildlife Fund due to the high number of species present in wetlands, the small global geographic area of wetlands, the number of species which are endemic to wetlands, and the high productivity of wetland systems. 100,000 species of animals alone 20,000 of them vertebrates, are living in wetland systems with 200 new species of freshwater fish being discovered at a rate of 200 a year.

Biodiverse River Basins: The Amazon holds 3,000 species of Freshwater fish species within the boundaries of its basin whose function it is to disperse the seeds of trees. *Brachyplatystoma vaillantii*, the Piramutaba catfish is migrates more than 3,300 km from its nursery grounds near the mouth of the Amazon River to its spawning grounds in Andean tributaries (400 m above sea level) distributing plants seed along the route.

Productive Intertidal Zones: Intertidal mudflats have a similar productivity even while possessing a low number of species. The abundance of invertebrates found within the mud are a food source for migratory waterfowl.

Critical Life-stage Habitat: Mudflats, saltmarshes, mangroves and seagrass beds contain bother species richness and productivity, home to important nursery areas for many commercial fish stocks.

Genetic Diversity: Many wetland systems are unique due to the long period of time that the ecosystem has been physically isolated from other aquatic sources. Lake Baikal in Russia has lake species found nowhere else in the world with immense biodiversity.

Lake Baikal: Evidence from a research study by Mazepova et. al suggest that the number of crustacean species endemic to Baikal Lake (>690 species and subspecies) exceeds the number of the same groups of animals inhabiting all freshwaters of Eurasia together. Its 150 species of free-living Platyhelminthes is analogous in number to the number of Platyhelminthes in all water bodies in Eastern Siberia. The species and subspecies number of Baikal sculpins (34) is more than twice the number of the analogous fauna inhabiting Eurasia. One of most exciting discoveries in this field was made by A.V. Shoshin who registered about 300 species of free-living nematodes using only 6 near-shore sampling localities in the Southern Baikal. If we will take into consideration, that about 60 per cent of the animals can be found nowhere else except Baikal, it may be assumed that the lake may be the biodiversity centre of the Eurasian continent.

Human Impact: Biodiversity loss occurs in wetland systems through land use changes, habitat destruction, pollution, exploitation of resources, and invasive species. Vulnerable, threatened, and endangered species number at 17 per cent of waterfowl, 38 per cent of fresh-water dependent mammals, 33 per cent of freshwater fish, 26 per cent of freshwater amphibians, 72 per cent of freshwater turtles, 86 per cent of marine turtles, 43 per cent of crocodilians and 27 per cent of coral reef-building species.

The impact of maintaining biodiversity is seen at the local level through job creation, sustainability, and community productivity. The Lower Mekong basin which runs through Cambodia, Laos, and Vietnam supports over 55 million people. Tourism is also enhanced through wildlife watching. The US state of Florida estimated US$ 1.6 billion was generated in state revenue from recreational activities associated with wildlife. Medicinal remedies have been found in native wetlands plants, such as the Red Mangrove *Rhizophora mangle* found in the Caribbean and Australia and known to possess antibacterial, wound-healing and anti-ulcer effects and antioxidant properties.

Wetland Products

Wetland systems naturally produce an array of vegetation and other ecological products that can harvested for personal and commercial use. The most significant of these is fish which have all or parts of their life-cycle occur in a wetland. Fresh and saltwater fish are the main source of protein for one billion people and comprise 15 per cent of an additional two billion people's diets. In addition, the fish generate the fishing industry in which over 80 per cent provides income and employment to residents in developing countries.

Another food staple found in wetland systems is rice a grain that is consumed at the rate of 1/5 of the total global calorie count. In Bangladesh, Cambodia and Vietnam, rice consumption reach 70 per cent with rice paddies predominant throughout Asia.

Rice is a food staple for 3 billion people worldwide and represents about one-fifth of global calorie consumption, rising to 70 per cent in some Asian countries such as Bangladesh, Cambodia and Vietnam.

Food converted to sweeteners and carbohydrates include the sago palm of Asia and Africa (cooking oil), the nipa palm of Asia (sugar, vinegar, alcohol, and fodder) and honey collection from mangroves. More than supplemental dietary intake, this produce sustains entire villages. Coastal Thailand villages earn the key portion of their income from sugar production while the country of Cuba relocates more than 30,000 hives to track the seasonal flowering of the mangrove *Avicennia*.

Other mangrove- derived products.

- fuelwood
- salt (produced by evaporating seawater)
- animal fodder
- traditional medicines (e.g. from mangrove bark)
- fibres for textiles
- dyes and tannins.

Human Impact: Over-fishing is the major problem for sustainable use of wetlands. Other issues that occur on a global level include contribution to climate change, point and non-point pollution, air and water quality issues.

The field of aquaculture within the fisheries industries is eliminating wetland through its destruction of mangroves especially with intensive shrimp farming. Aquaculture is developing throughout the Asia-Pacific region especially China with holdings of 90 per cent of the total quantity and 80 per cent of the value. Threats to rice fields mainly stem from inappropriate water management, introduction of invasive alien species, agricultural fertilizers and pesticides, and land use changes.

Industrial-scale production of palm oil threatens the biodiversity of wetland ecosystems in parts of south-east Asia, Africa and other developing country regions.

Exploitation of mangroves forests can occur in magnified instances such as the coastal villages of Southern Thailand who procure fuelwood and timber along with honey, resins, crab, and shellfish from these mangroves totalling around US$924 worth of wetland products annually per household.

Wetlands and Climate Change

Scientific Projections

"Low water and occasional drying of the wetland bottom during droughts (dry marsh phase) stimulate plant recruitment from a diverse seed bank and increase productivity by mobilising nutrients. In contrast, high water during deluges (lake marsh phase) causes turnover in plant populations and creates greater interspersion of element cover and open water, but lowers overall productivity. During a cover cycle that ranges from open water to complete vegetation cover, annual net primary productivity may vary 20-fold."

Mitigation and Adaption

Wetlands perform two important functions in relation to climate change. They have mitigation effects through their ability to sink carbon, and adaptation effects through their ability to store and regulate water.

Conservation

Wetlands have historically been the victim of large draining efforts for real estate development, or flooding for use as recreational lakes. By 1993 half the world's wetlands had been drained. Since the 1970s, more focus has been put on preserving wetlands for their natural function — sometimes also at great expense. Wetlands provide a valuable flood control function. Wetlands are very effective at filtering and cleaning water pollution, (often from agricultural runoff from the farms that replaced the wetlands in the first place). To replace these wetland ecosystem services enormous amounts of money have been spent on water purification plants and remediation measures, constructing dams, levees and other artificial flood controls.

In order to produce sustainable wetlands, short-term, private-sector profits need to come secondary to global equity. Decision-makers must valuate wetland type, provided ecosystem service, long-term benefit, and current subsidies inflating valuation on either the private or public sector side. Analysis using the impact of hurricanes versus storm protection features projected wetland valuation at US$33,000/hectare/year.

Balancing Wetland Conservation with the Needs of People

Wetlands are vital ecosystems that provide livelihoods for the millions of people who live in and around them. The Millennium Development Goals (MDGs) called for different sectors to join forces to secure wetland environments in the context of sustainable development and improving human wellbeing. A three-year project carried out by Wetlands International in partnership with the International Water Management Institute found that it is possible to conserve wetlands while improving the livelihoods of people living among them. Case studies conducted in Malawi and Zambia looked at how dambos – wet, grassy valleys or depressions where water seeps to the surface – can be farmed sustainably to improve livelihoods. Mismanaged or overused dambos often become degraded, however, using a knowledge exchange between local farmers and environmental managers, a protocol was developed using soil and water management practices. Project outcomes included a high yield of crops, development of sustainable farming techniques, and adequate water management generating enough water for use as irrigation. Before the project, there were cases where people had died from starvation due to food shortages. By the end of it, many more people had access to enough water to grow vegetables. A key achievement was that villagers had secure food supplies during long, dry months. They also benefited in other ways: nutrition was improved by growing a wider range of crops, and villagers could also invest in health and education by selling produce and saving money.

Ramsar Convention

The Convention on Wetlands of International Importance, especially as Waterfowl Habitat, or Ramsar Convention, is an international treaty designed to address global concerns regarding wetland loss and degradation. The primary purposes of the treaty are to list wetlands of international importance and to promote their wise use, with the ultimate goal of preserving the world's wetlands. Methods include restricting access to the majority portion of wetland areas, as well as educating the public to combat the misconception that wetlands are wastelands. The Convention works closely with five International Organisation Partners. These are: Birdlife International, IUCN, International Water Management Institute, Wetlands International and WWF. The partners provide technical expertise, help conduct or facilitate field studies and provide financial support. The IOPs also participate regularly as observers in all meetings of the Conference of the Parties and the Standing Committee and as full members of the Scientific and Technical Review Panel.

Conservation by Country

Conservation efforts vary in intensity and method by country.

New Zealand

Over 90 per cent of the wetlands in New Zealand have been drained since European settlement, predominantly to create farmland. Wetlands now have a degree of protection under the Resource Management Act.

South Africa

The South African Department of Environmental Affairs and Tourism in conjunction with the departments of Water Affairs and Forestry, and of Agriculture, supports the conservation and rehabilitation of wetlands through the Working for Wetlands programme. The aim of this programme is to encourage the protection, rehabilitation and sustainable use of South African wetlands through co-operative governance and partnerships. The programme is also a poverty relief effort, providing employment in wetland maintenance.

Sweden

The Swedish national wetland inventory (VMI) is one of the world's most extensive systematic inventories of nature types that has ever been done. VMI has surveyed the wetlands of Sweden below the alpine region during a 25-year period. In total 35 000 objects (sites) are included in VMI, corresponding to an area of 4.3 million hectares, or 10 per cent of the land area of Sweden. The aim of the survey has been to increase the general knowledge of wetlands in Sweden, as a basis for environmental monitoring and natural resources planning. By investigating the impact of human activities on wetlands and identifying the most valuable wetlands, their values can be preserved for future generations. The results from the inventory were also meant to function as background data for the authorities' decisions concerning e.g. drainage permits.

United States

The USA centred its wetland conservation efforts through legislation enforcement creating requirements that if a wetland had to be drained; developers had to offset the loss through the generation of artificial wetlands. A restoration project by the State of Florida in the Everglades acquired U.S. Sugar Corporation land allowing for water delivery, water treatment and water storage of sufficient quantity and quality to mimic the Everglades' natural system.

Republic of Macedonia

The fragments of wetland habitats that are still in existence in the Republic of Macedonia are present as marsh or swamp communities. These patches are present at Studenchishte (small fragment near Ohrid Lake), Pelagonia (village Chepigovo), Negortsi Spa, Bansko, Belchishte wetland and Monospitovo marsh. The large areas of swamps that used to be present in most of the valleys in contemporary Republic of Macedonia have undergone a great transformation over the last 50-60 years. The main cause for their reduction is land reclamation, drainage, and conversion into arable land for agricultural needs (Smith and Smith, 2003). Some of the remaining wetlands (Negortsi Spa, Bansko) are of great importance for understanding the genesis of marsh vegetation in the Republic of Macedonia as many mountainous marshes and peat bogs suffered anthropogenic transformation due to the capturing of water from mountain springs and streams for the purposes of generating drinking water (Smith, 2003). Accordingly, the fragmentation and transformation of previous swamps had a major impact on faunal distribution and abundance. Amphibians are the most affected species along with invertebrate and vertebrate groups including the European Otter (Lutra Lutra L.). It is listed as Nearly Threatened according to the IUCN's red list and mainly found now only in the Belchishte wetland. The otter has importance for the wetland communities not just in Macedonia but also in other European countries such as the Netherlands and Germany. Macedonian wetlands lie within the network of some of the bigger rivers in Macedonia including the Vardar River, a catchment area equal to 80 per cent of Macedonian territory. .

Valuation

The value of a wetland system to the earth and to humankind is one of the most important valuations that can be computed for sustainable development. A guideline involving assessing a wetland, keeping inventories of known wetlands, and monitoring the same wetlands over time is the current process that is used to educate environmental decision-makers such as governments on the importance of wetland protection and conservation.

Important Considerations

- Constructed Wetlands take 10-100 years to fully resemble the vegetative composition of a natural wetland.
- Artificial wetlands do not have hydric soil. The soil has very low levels of organic carbon and total nitrogen compared to natural wetland systems.
- Organic matter can be added to degraded natural wetlands to help restore their productivity before the wetland is destroyed.

Assessment

Rapid assessment.

Five steps to assessing a wetland;

1. Collect general biodiversity data in order to inventory and prioritise wetland species, communities and ecosystems. Obtain baseline biodiversity information for a given area.
2. Gather information on the status of a focus or target species (such as threatened species. Collect data pertaining to the conservation of a specific species.
3. Gain information on the effects of human or natural disturbance (changes) on a given area or species.
4. Gather information that is indicative of the general ecosystem health or condition of a specific wetland ecosystem.
5. Determine the potential for sustainable use of biological resources in a particular wetland ecosystem.

Inventory

Developing a global inventory of wetlands has proven to be a large and difficult undertaking. Current efforts are based on available data, but both classification and spatial resolution have proven to be inadequate for regional or site-specific environmental management decision-making. It is difficult to identify small, long, and narrow wetlands within the landscape.

Many of today's remote sensing satellites do not have sufficient spatial and spectral resolution to monitor wetland conditions, although multispectral IKONOS and QuickBird data may offer improved spatial resolutions once it is 4 m or higher. Majority of the pixels are just mixtures of several plant species or vegetation types and are difficult to isolate which translates into an inability to classify the vegetation that defines the wetland.

Improved remote sensing information, coupled with good knowledge domain on wetlands will facilitate expanded efforts in wetland monitoring and mapping. This will also be extremely important because we expect to see major shifts in species composition due to both anthropogenic land use and natural changes in the environment caused by climate change.

Monitoring

Mapping

A wetland system needs to be monitored over time to in order to assess whether it is functioning at an ecologically sustainable level or whether it is becoming degraded. Degraded wetlands will suffer a loss in water quality, a high number of threatened and endangered species, and poor soil conditions.

Due to the large size of wetlands, mapping is an effective tool to monitor wetlands. There are many remote sensing methods that can be used to map wetlands. Remote-sensing technology permits the acquisition of timely digital data on a repetitive basis. This repeat coverage allows wetlands, as well as the adjacent land-cover and land-use types, to be monitored seasonally and/ or annually. Using digital data provides a standardised data-collection procedure and an opportunity for data integration within a geographic information system. Traditionally, Landsat 5 Thematic Mapper (TM), Landsat 7 Enhanced Thematic Mapper Plus (ETM +), and the SPOT 4 and 5 satellite systems have been used for this purpose. More recently, however, multispectral IKONOS and QuickBird data, with spatial resolutions of 4 m by 4 m and 2.44 m by 2.44 m, respectively, have been shown to be excellent sources of data when mapping and monitoring smaller wetland habitats and vegetation communities.

For example, Detroit Lakes Wetland Management District assessed area wetlands in Michigan, USA using remote sensing. Through using this technology, satellite images were taken over a large geographic area and extended period. In addition, using this technique was less costly and time-consuming compared to the older method using visual interpretation of aerial photographs. In comparison, most aerial photographs also require experienced interpreters to extract information based on structure and texture while the interpretation of remote sensing data only requires analysis of one characteristic (spectral). However, there are a number of limitations associated with this type of image acquisition. Analysis of wetlands has proved difficult because to obtain the data because it is often linked to other purposes such as the analysis of land cover or land use. Practically, many natural wetlands are difficult to monitor as these areas are quite often difficult to access and require exposure to native wildlife and potential endemic disease.

Further Improvements

Methods to develop a classification system for specific biota of interest could assist with technological advances that will allow for identification at a very high accuracy rate. The issue of the cost and expertise involved in remote sensing technology is still a factor hindering further advancements in image acquisition and data processing. Future improvements in current wetland vegetation mapping could include the use of more recent and better geospatial data when it is available.

List of Wetland Types

Wetland types:

A Marine and Coastal Zone wetlands

1. Marine waters—permanent shallow waters less than six metres deep at low tide; includes sea bays, straits
2. Subtidal aquatic beds; includes kelp beds, seagrasses, tropical marine meadows
3. Coral reefs
4. Rocky marine shores; includes rocky offshore islands, sea cliffs
5. Sand, shingle or pebble beaches; includes sand bars, spits, sandy islets
6. Intertidal mud, sand or salt flats
7. Intertidal marshes; includes saltmarshes, salt meadows, saltings, raised salt marshes, tidal brackish and freshwater marshes
8. Intertidal forested wetlands; includes mangrove swamps, nipa swamps, tidal freshwater swamp forests
9. Brackish to saline lagoons and marshes with one or more relatively narrow connections with the sea

10. Freshwater lagoons and marshes in the coastal zone
11. Non-tidal freshwater forested wetlands

B Inland wetlands

1. Permanent rivers and streams; includes waterfalls
2. Seasonal and irregular rivers and streams
3. Inland deltas (permanent)
4. Riverine floodplains; includes river flats, flooded river basins, seasonally flooded grassland, savanna and palm savanna
5. Permanent freshwater lakes (> 8 ha); includes large oxbow lakes
6. Seasonal/intermittent freshwater lakes (> 8 ha), floodplain lakes
7. Permanent saline/brackish lakes
8. Seasonal/intermittent saline lakes
9. Permanent freshwater ponds (< 8 ha), marshes and swamps on inorganic soils; with emergent vegetation waterlogged for at least most of the growing season
10. Seasonal/intermittent freshwater ponds and marshes on inorganic soils; includes sloughs, potholes; seasonally flooded meadows, sedge marshes
11. Permanent saline/brackish marshes
12. Seasonal saline marshes
13. Shrub swamps; shrub-dominated freshwater marsh, shrub carr, alder thicket on inorganic soils
14. Freshwater swamp forest; seasonally flooded forest, wooded swamps; on inorganic soils
15. Peatlands; forest, shrub or open bogs
16. Alpine and tundra wetlands; includes alpine meadows, tundra pools, temporary waters from snow melt
17. Freshwater springs, oases and rock pools
18. Geothermal wetlands
19. Inland, subterranean karst wetlands

C Human-made wetlands

1. Water storage areas; reservoirs, barrages, hydro-electric dams, impoundments (generally > 8 ha)
2. Ponds, including farm ponds, stock ponds, small tanks (generally < 8 ha)

3. Aquaculture ponds; fish ponds, shrimp ponds
4. Salt exploitation; salt pans, salines
5. Excavations; gravel pits, borrow pits, mining pools
6. Wastewater treatment; sewage farms, settling ponds, oxidation basins
7. Irrigated land and irrigation channels; rice fields, canals, ditches
8. Seasonally flooded arable land, farm land
9. Canals.

Variations of names for wetland systems:

- Swamp
  - Freshwater swamp forest
  - Coniferous swamp
- Peat swamp forest
- Slough
- Flooded grasslands and savannas
- Constructed wetland
- Riparian zone
- Fen
- Vernal Pool.

Ocean

An ocean is a major body of saline water, and a principal component of the hydrosphere. Approximately 71 per cent of the Earth's surface (~3.6×10^8 km^2) is covered by ocean, a continuous body of water that is customarily divided into several principal oceans and smaller seas.

More than half of this area is over 3,000 metres (9,800 ft) deep. Average oceanic salinity is around 35 parts per thousand (‰) (3.5 per cent), and nearly all seawater has a salinity in the range of 30 to 38 ‰. Scientists estimate that 230,000 marine species are currently known, but the total could be up to 10 times that number.

Though generally described as several 'separate' oceans, these waters comprise one global, interconnected body of salt water sometimes referred to as the World Ocean or global ocean. This concept of a continuous body of water with relatively free interchange among its parts is of fundamental importance to oceanography. The major oceanic divisions are defined in part by the continents, various archipelagos, and other criteria. These divisions are (in descending order of size):

- Pacific Ocean, which separates Asia and Australia from the Americas
- Atlantic Ocean, which separates the Americas from Europe and Africa
- Indian Ocean, which washes upon southern Asia and separates Africa and Australia
- Southern Ocean, sometimes considered an extension of the Pacific, Atlantic and Indian Oceans, which encircles Antarctica.
- Arctic Ocean, sometimes considered a sea of the Atlantic, which covers much of the Arctic and washes upon northern North America and Eurasia.

The Pacific and Atlantic may be further subdivided by the equator into northern and southern portions. Smaller regions of the oceans are called seas, gulfs, bays, straits and other names.

Geologically, an ocean is an area of oceanic crust covered by water. Oceanic crust is the thin layer of solidified volcanic basalt that covers the Earth's mantle. Continental crust is thicker but less dense. From this perspective, the earth has three oceans: the World Ocean, the Caspian Sea, and Black Sea. The latter two were formed by the collision of Cimmeria with Laurasia. The Mediterranean Sea is at times a discrete ocean, because tectonic plate movement has repeatedly broken its connection to the World Ocean through the Strait of Gibraltar. The Black Sea is connected to the Mediterranean through the Bosporus, but the Bosporus is a natural canal cut through continental rock some 7,000 years ago, rather than a piece of oceanic sea floor like the Strait of Gibraltar. Despite their names, smaller landlocked bodies of saltwater that are *not* connected with the World Ocean, such as the Aral Sea, are actually salt lakes.

Borders of the Oceans

Ocean and Life: The ocean has a significant effect on the biosphere. Oceanic evaporation, as a phase of the water cycle, is the source of most rainfall, and ocean temperatures determine climate and wind patterns that affect life on land. Life within the ocean evolved 3 billion years prior to life on land. Both the depth and distance from shore strongly influence the amount and kinds of plants and animals that live there.

Physical Properties

The area of the World Ocean is 361 million square kilometres (139 million square miles) Its volume is approximately 1.3 billion cubic kilometres (310 million cu mi). This can be thought of as a cube of water with an edge length of 1,111 kilometres (690 mi). Its average depth is 3,790 metres (12,430 ft), and its maximum depth is 10,923 metres (6.787 mi) Nearly half of the world's marine waters are over 3,000 metres (9,800 ft) deep. The vast expanses of deep ocean (anything below 200 metres (660 ft)) cover about 66 per cent of the Earth's surface. This does not include seas not connected to the World Ocean, such as the Caspian Sea.

The total mass of the hydrosphere is about 1,400,000,000,000,000,000 metric tons (1.5×10^{18} short tons) or 1.4×10^{21} kg, which is about 0.023 percent of the Earth's total mass. Less than 3 percent is freshwater; the rest is saltwater, mostly in the ocean.

Colour

A common misconception is that the oceans are blue primarily because the sky is blue. In fact, water has a very slight blue colour that can only be seen in large volumes. While the sky's reflection does contribute to the blue appearance of the surface, it is not the primary cause. The primary cause is the absorption by the water molecules of red photons from the incoming light, the only known example of colour in nature resulting from vibrational, rather than electronic, dynamics.

Glow

Sailors and other mariners have reported that the ocean often emits a visible glow, or luminescence, which extends for miles at night. In 2005, scientists announced that for the first time, they had obtained photographic evidence of this glow. It may be caused by bioluminescence.

Exploration

The deepest point in the ocean is the Mariana Trench, located in the Pacific Ocean near the Northern Mariana Islands. Its maximum depth has been estimated to be 10,971 metres (35,994 ft) (plus or minus 11 meters; see the Mariana Trench article for discussion of the various estimates of the maximum depth.) The British naval vessel, *Challenger II* surveyed the trench in 1951 and named the deepest part of the trench, the "Challenger Deep". In 1960, the Trieste successfully reached the bottom of the trench, manned by a crew of two men.

Much of the ocean bottom remains unexplored and unmapped. A global image of many underwater features larger than 10 kilometres (6.2 mi) was created in 1995 based on gravitational distortions of the nearby sea surface.

Regions and Depths

Oceanographers divide the ocean into regions depending on physical and biological conditions of these areas. The pelagic zone includes all open ocean regions, and can be divided into further regions categorised by depth and light abundance. The photic zone covers the oceans from surface level to 200 metres down. This is the region where photosynthesis can occur and therefore is the most biodiverse. Since plants require photosynthesis, life found deeper than this must either rely on material sinking from above or find another energy source; hydrothermal vents are the primary option in what is known as the aphotic zone (depths exceeding 200 m). The pelagic part of the photic zone is known as the epipelagic. The pelagic part of the aphotic zone can be further divided into regions that succeed each other vertically according to temperature.

The mesopelagic is the uppermost region. Its lowermost boundary is at a thermocline of 12°C (54 °F), which, in the tropics generally lies at 700–1,000 metres (2,300–3,300 ft). Next is the bathypelagic lying between 10 and 4°C (50 and 39 °F), typically between 700–1,000 metres (2,300–3,300 ft) and 2,000–4,000 metres (6,600–13,000 ft) Lying along the top of the abyssal plain is the abyssalpelagic, whose lower boundary lies at about 6,000 metres (20,000 ft). The last zone includes the deep trenches, and is known as the hadalpelagic. This lies between 6,000–11,000 metres (20,000–36,000 ft) and is the deepest oceanic zone.

Along with pelagic aphotic zones there are also benthic aphotic zones. These correspond to the three deepest zones of the deep-sea. The bathyal zone covers the continental slope down to about 4,000 metres (13,000 ft). The abyssal zone covers the abyssal plains between 4,000 and 6,000 m. Lastly, the hadal zone corresponds to the hadalpelagic zone which is found in the oceanic trenches.

The pelagic zone can also be split into two subregions, the neritic zone and the oceanic zone. The neritic encompasses the water mass directly above the continental shelves, while the oceanic zone includes all the completely open water. In contrast, the littoral zone covers the region between low and high tide and represents the transitional area between marine and terrestrial conditions. It is also known as the intertidal zone because it is the area where tide level affects the conditions of the region.

Geology

The ocean floor spreads from mid-ocean ridges where two plates adjoin. Where two plates move towards each other, one plate subducts under another plate (oceanic or continental) leading to an oceanic trench.

Climate Effects

Ocean currents greatly affect the Earth's climate by transferring heat from the tropics to the polar regions, and transferring warm or cold air and precipitation to coastal regions, where winds may carry them inland. Surface heat and freshwater fluxes create global density gradients that drive the thermohaline circulation part of large-scale ocean circulation. It plays an important role in supplying heat to the polar regions, and thus in sea ice regulation. Changes in the thermohaline circulation are thought to have significant impacts on the Earth's radiation budget. Insofar as the thermohaline circulation governs the rate at which deep waters reach the surface, it may also significantly influence atmospheric carbon dioxide concentrations.

It is often stated that the thermohaline circulation is the primary reason that the climate of Western Europe is so temperate. An alternate hypothesis claims that this is largely incorrect, and that Europe is warm mostly because it lies downwind of an ocean basin, and because atmospheric waves bring warm air north from the subtropics. The Antarctic Circumpolar Current encircles that continent, influencing the area's climate and connecting currents in several oceans. One of the most dramatic forms of weather occurs over the oceans: tropical cyclones (also called "typhoons" and "hurricanes" depending upon where the system forms).

Biology

Lifeforms native to oceans include:

- Radiata
- Fish
- Cetacea such as whales, dolphins and porpoises,
- Cephalopods such as octopus and squid
- Crustaceans such as lobsters, clams, shrimp and krill
- Marine worms
- Plankton
- Echinoderms such as brittle stars, starfish, sea cucumbers and sand dollars.

Economy

The oceans are essential to transportation: most of the world's goods move by ship between the world's seaports.

Oceans are also the major supply source for the fishing industry. Some of the more major ones are shrimp, fish, crabs and lobster.

Ancient Oceans

Continental drift continually reconfigures the oceans, joining and splitting bodies of water. Ancient oceans include:

- Bridge River Ocean, the ocean between the ancient Insular Islands and North America.
- Iapetus Ocean, the southern hemisphere ocean between Baltica and Avalonia.
- Panthalassa, the vast world ocean that surrounded the Pangaea supercontinent.

- Rheic Ocean
- Slide Mountain Ocean, the ocean between the ancient Intermontane Islands and North America.
- Tethys Ocean, the ocean between the ancient continents of Gondwana and Laurasia.
- Khanty Ocean, the ocean between Baltica and Siberia.
- Mirovia, the ocean that surrounded the Rodinia supercontinent.
- Paleo-Tethys Ocean, the ocean between Gondwana and the Hunic terranes.
- Poseidon Ocean
- Proto-Tethys Ocean
- Pan-African Ocean, the ocean that surrounded the Pannotia supercontinent.
- Superocean, the ocean that surrounds a global supercontinent.
- Ural Ocean, the ocean between Siberia and Baltica.

Extraterrestrial Oceans

Earth is the only known planet with liquid water on its surface and is certainly the only one in our own solar system. However, a layer of liquid water thick enough to decouple the crust from the mantle is thought to be present under the surfaces of the moons Titan, Europa and, with less certainty, Callisto and Ganymede. A similar magma ocean is thought to be present on Io. Geysers have been found on Saturn's moon Enceladus, though these may not involve bodies of liquid water. Other icy moons and trans-neptunian objects may also have internal oceans, or have once had internal oceans that have now frozen. The planets Uranus and Neptune may also possess large oceans of liquid water under their thick atmospheres, though their internal structure is not well understood.

There is currently much debate over whether Mars once had an ocean in its northern hemisphere, and over what happened to it; recent findings by the Mars Exploration Rover mission indicate Mars had long-term standing water in at least one location, but its extent is not known.

Astronomers believe that Venus had liquid water and perhaps oceans in its very early history. If they existed, all later vanished via resurfacing.

Liquid hydrocarbons (not water) are thought to be present on the surface of Titan, though *lakes* may be a more accurate term. The Cassini–Huygens space mission initially discovered only what appeared to be dry lakebeds and empty river channels, suggesting that Titan had lost what surface liquids it might have had. Cassini's more recent fly-by of Titan offers radar images that strongly suggest hydrocarbon lakes near the colder polar regions. Titan is thought to have a subterranean water ocean under the ice and hydrocarbon mix that forms its outer crust.

Beyond the solar system, the planet Gliese 581 c is at the right distance from its sun to support liquid surface water. However, its greenhouse effect would make it too hot for oceans to exist on the surface. On Gliese 581 d the greenhouse effect may bring temperatures suitable for surface oceans. Astronomers dispute whether HD 209458 b has water vapour in its atmosphere. Gliese 436 b is believed to have "hot ice." Neither of these planets are cool enough for liquid water—but if water molecules exist there, they are also likely to be found on planets at a suitable temperature. GJ 1214 b, detected by transit, found evidence that this planet has oceans made of exotic form of ice VII, making up 75 per cent of all the planet's mass.

Culture

The original concept of "ocean" goes back to notions of Mesopotamian and Indo-European mythology, imagining the world to be encircled by a great river. Okeanos in Greek, reflects the ancient Greek observation that a strong current flowed off Gibraltar and their subsequent assumption that it was a great river. (Compare also *Samudra* from Hindu mythology and Jörmungandr from Norse mythology.) The world was imagined to be enclosed by a celestial ocean above the heavens, and an ocean of the underworld below. Artworks which depict maritime themes are known as marine art, a term which particularly applies to common styles of European painting of the 17th to 19th centuries.

Evapotranspiration

Evapotranspiration (ET) is a term used to describe the sum of evaporation and plant transpiration from the Earth's land surface to atmosphere. Evaporation accounts for the movement of water to the air from sources such as the soil, canopy interception, and waterbodies. Transpiration accounts for the movement of water within a plant and the subsequent loss of water as vapour through stomata in its leaves. Evapotranspiration is an important part of the water cycle. An element (such as a tree) that contributes to evapotranspiration can be called an evapotranspirator.

Potential evapotranspiration (PET) is a representation of the environmental demand for evapotranspiration and represents the evapotranspiration rate of a short green crop, completely shading the ground, of uniform height and with adequate water status in the soil profile. It is a reflection of the energy available to evaporate water, and of the wind available to transport the water vapour from the ground up into the lower atmosphere. Evapotranspiration is said to equal potential evapotranspiration when there is ample water.

Evapotranspiration and the Water Cycle

Evapotranspiration is a significant water loss from drainage basins. Types of vegetation and land use significantly affect evapotranspiration, and therefore the amount of water leaving a drainage basin. Because water transpired through leaves comes from the roots, plants with deep reaching roots can more constantly transpire water. Herbaceous plants generally transpire less than woody plants because they usually have less extensive foliage. Conifer forests tend to have higher rates of evapotranspiration than deciduous forests, particularly in the dormant and early spring seasons. This is primarily due to the enhanced amount of precipitation intercepted and evaporated by conifer foliage during these periods. Factors that affect evapotranspiration include the plant's growth stage or level of maturity, percentage of soil cover, solar radiation, humidity, temperature, and wind.

Through evapotranspiration, forests reduce water yield, except in unique ecosystems called cloud forests. Trees in cloud forests condense fog or low clouds into liquid water on their surface, which drips down to the ground. These trees still contribute to evapotranspiration, but often condense more water than they evaporate or transpire. In areas that are not irrigated, actual evapotranspiration is usually no greater than precipitation, with some buffer in time depending on the soil's ability to hold water. It will usually be less because some water will be lost due to percolation or surface runoff. An exception is areas with high water tables, where capillary action can cause water from the groundwater to rise through the soil matrix to the surface. If potential evapotranspiration is greater than actual precipitation, then soil will dry out, unless irrigation is used. Evapotranspiration can never be greater than PET, but can be lower if there is not enough water to be evaporated or plants are unable to transpire readily.

Estimating Evapotranspiration

Evapotranspiration can be measured or estimated using several methods.

Indirect Methods

Pan evaporation data can be used to estimate lake evaporation, but transpiration and evaporation of intercepted rain on vegetation are unknown. There are three general approaches to estimate evapotranspiration indirectly.

Catchment Water Balance

Evapotranspiration may be estimated by creating an equation of the water balance of a drainage basin. The equation balances the change in water stored within the basin (S) with inputs and exports:

$$\Delta S = P - ET - Q - D$$

The input is precipitation (P), and the exports are evapotranspiration (which is to be estimated), streamflow (Q), and groundwater recharge (D). If the change in storage, precipitation, streamflow, and groundwater recharge are all estimated, the missing flux, ET, can be estimated by rearranging the above equation as follows:

$$ET = P - \Delta S - Q - D$$

Hydrometeorological Equations

The most general and widely used equation for calculating reference ET is the Penman equation. The Penman-Monteith variation is recommended by the Food and Agriculture Organisation. The simpler Blaney-Criddle equation was popular in the Western United States for many years but it is not as accurate in regions with higher humidities. Other solutions used includes Makkink, which is simple but must be calibrated to a specific location, and Hargreaves. To convert the reference evapotranspiration to actual crop evapotranspiration, a crop coefficient and a stress coefficient must be used. Crop coefficients referred to in many hydrological models are themselves the result of equations that describe predictable variation in coefficient values depending upon plant conditions that change during periods for which the model is used. This is because crops are seasonal, perennial plants mature over multiple seasons, and stress responses can significantly depend upon many aspects of plant condition.

Energy Balance

A third methodology to estimate the actual evapotranspiration is the use of the energy balance.

$$\lambda E = R_n + G - H$$

where λE is the energy needed to change the phase of water from liquid to gas, R_n is the net radiation, G is the soil heat flux and H is the sensible heat flux. Using instruments like a scintillometer, soil heat flux plates or radiation meters, the components of the energy balance can be calculated and the energy available for actual evapotranspiration can be solved.

The SEBAL algorithm solves the energy balance at the earth surface using satellite imagery. This allows for both actual and potential evapotranspiration to be calculated on a pixel-by-pixel basis. Evapotranspiration is a key indicator for water management and irrigation performance. SEBAL can map these key indicators in time and space, for days, weeks or years.

Self-Organising Map Estimator

As mentioned previously, Reference crop evapotranspiration (ETo) estimation is of importance in irrigation water management for the calculation of crop water requirements and its scheduling, in rainfall-runoff modelling and in numerous other water resources studies.

Due to its importance, several direct and indirect methods have been employed to determine the reference crop evapotranspiration but success has been limited because the direct measurement methods lack in precision and accuracy due to scale issues and other problems, while some of the more accurate indirect methods, e.g. the Penman-Monteith benchmark model, are time-consuming and require weather input data that are not routinely monitored.

Adeloye et al. (2011) used the Kohonen Self-Organising Map (KSOM), unsupervised artificial neural networks, to predict the ETo. based on observed daily weather data at two climatically diverse basins: a small experimental catchment in temperate Edinburgh, UK and a semi-arid lake basin in Udaipur, India.

This was achieved by using the powerful clustering capability of the KSOM to analyse the multi-dimensional data array comprising the estimated ETo (based on the FAO Penman-Monteith model) and different subsets of climatic variables known to affect it.

The findings indicate that the KSOM-based ETo estimates even with fewer input variables were in good agreement with those obtained using the conventional FAO Penman-Monteith formulation employing the full complement of weather data at the two locations.

More crucially, the KSOM-based estimates were also found to be significantly superior to those estimated using currently recommended empirical ETo methods for data scarce situations such as those in developing countries.

Experimental Method for Measuring ET

One method for measuring ET is with a weighing lysimeter. The weight of a soil column is measured continuously and the change in storage of water in the soil is modelled by the change in weight. The change in weight is converted to units of length based on the surface area of the weighing lysimeter and the unit weight of water. ET is computed as the change in weight plus rainfall minus percolation.

Eddy Covariance

The most direct method of measuring evapotranspiration is with the eddy covariance technique in which fast fluctuations of vertical wind speed are correlated with fast fluctuations in atmospheric water vapor density. This directly estimates the transfer of water vapor (evapotranspiration) from the land (or canopy) surface to the atmosphere.

Potential Evapotranspiration

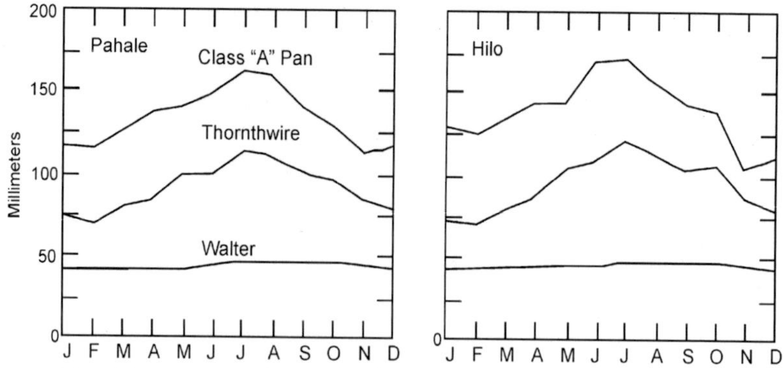

Figure: *Monthly estimated potential evapotranspiration and measured pan evaporation for two locations in Hawaii, Hilo and Pahala.*

Potential evapotranspiration (PET) is the amount of water that would be evaporated and transpired if there were sufficient water available. This demand incorporates the energy available for evaporation and the ability of the lower atmosphere to transport evaporated moisture away from the land surface. PET is higher in the summer, on less cloudy days, and closer to the equator, because of the higher levels of solar radiation that provides the energy for evaporation. PET is also higher on windy days because the evaporated moisture can be quickly moved from the ground or plant surface, allowing more evaporation to fill its place. PET is expressed in terms of a depth of water, and can be graphed during the year.

Potential evapotranspiration is usually measured indirectly, from other climatic factors, but also depends on the surface type, such as free water (for lakes and oceans), the soil type for bare soil, and the vegetation. Often a value for the potential evapotranspiration is calculated at a nearby climate station on a reference surface, conventionally short grass. This value is called the reference evapotranspiration, and can be converted to a potential evapotranspiration by multiplying with a surface coefficient. In agriculture, this is called a crop coefficient. The difference between potential evapotranspiration and precipitation is used in irrigation scheduling. Average annual PET is often compared to average annual precipitation, P. The ratio of the two, P/PET, is the aridity index.

Reservoir

A reservoir or an artificial lake is used to store water. Reservoirs may be created in river valleys by the construction of a dam or may be built by excavation in the ground or by conventional construction techniques such as brickwork or cast concrete. The term reservoir may also be used to describe underground reservoirs such as an oil or water well.

Types

Valley Dammed Reservoir

A dam constructed in a valley relies on the natural topography to provide most of the basin of the reservoir. Dams are typically located at a narrow part of a valley downstream of a natural basin. The valley sides act as natural walls with the dam located at the narrowest practical point to provide strength and the lowest practical cost of construction. In many reservoir construction projects people have to be moved and re-housed, historical artifacts moved or rare environments relocated. Examples include the temples of Abu Simbel (which were moved before the construction of the Aswan Dam to create Lake Nasser from the Nile in Egypt) and the re-location of the village of Capel Celyn during the construction of Llyn Celyn.

Construction of a reservoir in a valley will usually necessitate the diversion of the river during part of the build often through a temporary tunnel or by-pass channel.

In hilly regions reservoirs are often constructed by enlarging existing lakes. Sometimes in such reservoirs the new top water level exceeds the watershed height on one or more of the feeder streams such as at Llyn Clywedog in Mid Wales. In such cases additional side dams are required to contain the reservoir.

Where the topography is poorly suited to a single large reservoir, a number of smaller reservoirs may be constructed in a chain such as in the River Taff valley where the three reservoirs Llwyn-on Reservoir, Cantref Reservoir and Beacons Reservoir form a chain up the valley.

Bank-side Reservoir

Where water is taken from a river of variable quality or quantity, bank-side reservoirs may be constructed to store the water pumped or siphoned from the river. Such reservoirs are usually built partly by excavation and partly by the construction of a complete encircling bund or embankment which may exceed 6 km in circumference. Both the floor of the reservoir and the bund must have an impermeable lining or core, often made of puddled clay. The water stored in such reservoirs may have a residence time of several months during which time normal biological processes are able to substantially reduce many contaminants and almost eliminate any turbidity. The use of bank-side reservoirs also allows a water abstraction to be closed down for extended period at times when the river is unacceptably polluted or when flow conditions are very low due to drought. The London water supply system is one example of the use of bank-side storage for all the water taken from the River Thames and River Lee with many large reservoirs such as Queen Mary Reservoir visible along the approach to London Heathrow Airport.

Service Reservoir

Service reservoirs store fully treated potable water close to the point of distribution. Many service reservoirs are constructed as water towers, often as elevated structures on concrete pillars where the landscape is relatively flat. Other service reservoirs are entirely underground, especially in more hilly or mountainous country. In the United Kingdom, Thames Water has many underground reservoirs built in the 1800s by the Victorians, most of which are lined with brick. A good example is the Honour Oak Reservoir, constructed between 1901 and 1909. When it was completed it was the largest brick built underground reservoir in the world and is still one of the largest in Europe. The reservoir now forms part of the Southern extension of the Thames Water Ring Main. The top of the reservoir has been grassed over and is now the Aquarias Golf Club.

Service reservoirs perform several functions including ensuring sufficient head of water in the water distribution system and providing hydraulic capacitance in the system to even out peak demand from consumers enabling the treatment plant to run at optimum efficiency. Large service reservoirs can also be managed to so that energy costs in pumping are reduced by concentrating refilling activity at times of day when power costs are low.

History

Five thousand years ago, the craters of extinct volcanoes in Arabia were used as reservoirs by farmers for their irrigation water.

Dry climate and water scarcity in India led to early development of water management techniques, including the building of a reservoir at Girnar in 3000 BC. Artificial lakes dating to the 5th century BC have been found in ancient Greece. An artificial lake in present-day Madhya Pradesh province of India, constructed in the 11th century, covered 650 square metres (7,000 sq ft).

In Sri Lanka large reservoirs have been created by ancient Sinhalese kings in order to save the water for irrigation. The famous Sri Lankan king Parakramabahu I of Sri Lanka stated " do not let a drop of water seep into the ocean without benefiting mankind ". He created the reservoir named Parakrama Samudra(sea of King Parakrama), which has astonished archaeologists.

Uses

Direct Water Supply

Many dammed river reservoirs and most bank-side reservoirs are used to provide the raw water feed to a water treatment plant which delivers drinking water through water mains. The reservoir does not simply hold water until it is needed; it can also be the first part of the water treatment process.

The time the water is held for before it is released is known as the *retention time*. This is a design feature that allows particles and silts to settle out, as well as time for natural biological treatment using algae, bacteria and zooplankton that naturally live within the water. However natural limnological processes in temperate climate lakes produces temperature stratification in the water body which tends to partition some elements such as manganese and phosphorus into deep, cold anoxic water during the summer months. In the autumn and winter the lake becomes fully mixed again. During drought conditions, it is sometimes necessary to draw down the cold bottom water and the elevated levels of manganese in particular can cause problems in water treatment plants.

Hydroelectricity

A reservoir generating hydroelectricity includes turbines connected to the retained water body by large-diameter pipes. These generating sets may be at the base of the dam or some distance away. Some reservoirs generating hydroelectricity use pumped re-charge in which a high-level reservoir is filled with water using high-performance electric pumps at times when electricity demand is low and then uses this stored water to generate electricity by releasing the stored water into a low-level reservoir when electricity demand is high. Such systems are called pump-storage schemes.

Controlling Watercourses

Reservoirs can be used in a number of ways to control how water flows through downstream waterways.

Downstream water supply – water may be released from an upland reservoir so that it can be abstracted for drinking water lower down the system, sometimes hundred of miles further down downstream

Irrigation – water in an irrigation reservoir may be released into networks of canals for use in farmlands or secondary water systems. Irrigation may also be supported by reservoirs which maintain river flows allowing water to be abstracted for irrigation lower down the river.

Flood control – also known as an "*attenuation*" or "*balancing*" reservoir, flood control reservoirs collect water at times of very high rainfall, then release it slowly over the course of the following weeks or months. Some of these reservoirs are constructed across the river line with the onward flow controlled by an orifice plate. When river flow exceeds the capacity of the orifice plate water builds behind the dam but as soon as the flow rate reduces the water behind the dam slowly releases until the reservoir is empty again. In some cases such reservoirs only function a few times in a decade and the land behind the reservoir may be developed as community or recreational land. A new generation of balancing dams are being developed to combat the climatic consequences of climate change. They are called “Flood Detention Reservoirs”. Because these reservoirs will remain dry for long periods, there may be a risk of the clay core drying out reducing its structural stability. Recent developments include the use of composite core fill made from recycled materials as an alternative to clay.

Canals – Where a natural watercourse’s water is not available to be diverted into a canal, a reservoir may be built to guarantee the water level in the canal; for example, where a canal climbs to cross a range of hills through locks.

Recreation – water may be released from a reservoir to artificially create or supplement white-water conditions for kayaking and other white-water sports. On salmonid rivers special releases (in Britain called *freshets*) are made to encourage natural migration behaviours in fish and to provide a variety of fishing conditions for anglers.

Flow Balancing

Reservoirs can be used to balance the flow in highly managed systems, taking in water during high flows and releasing it again during low flows. In order for this to work without pumping requires careful control of water levels using adjustable sluices. Accurate weather forecasts are essential so that dam operators can plan drawdowns prior to a high rainfall or snowmelt event. Dam operators blamed a faulty weather forecast on the 2010–2011 Queensland floods. Examples of highly managed Reservoirs are Burrendong Dam in Australia and Llyn Tegid in North Wales. Llyn Tegid is a natural lake whose level was raised by a low dam and into which the River Dee flows or discharges depending upon flow conditions at the time as part of the River Dee regulation system. This mode of operation is a form of hydraulic capacitance in the river system.

Recreation

The water bodies provided by many reservoirs often allow some recreational uses such as fishing, boating, and other activities. Special rules may apply for the safety of the public and to protect the quality of the water and the ecology of the surrounding area. Many reservoirs now support and encourage less informal and less structured recreation such as natural history, bird watching, landscape painting, walking and hiking and often provide information boards and interpretation material to encourage responsible use.

Operation

Water falling as rain upstream of the reservoir together with any groundwater emerging as springs is stored in the reservoir. Any excess water can be spilled via a specifically designed spillway. Stored water may be piped by gravity for use as drinking water, to generate hydro-electricity or to maintain river flows to support downstream uses. Occasionally reservoirs can be managed to retain high rain-fall events to prevent or reduce downstream flooding. Some reservoirs support several uses and the operating rules may be complex.

Most modern reservoirs have a specially designed draw-off tower that can discharge water from the reservoir at different levels both to access water as the reservoir draws down but also to allow water of a specific quality to be discharged into the downstream river as compensation water.

The operators of many upland or in-river reservoirs have obligations to release water into the downstream river to maintain river quality, support fisheries, maintain downstream industrial uses. Maintain recreational use or for a range of other requirements. Such releases are known as *compensation water*.

Terminology

The terminology for reservoirs varies from country to country. In most of the world reservoir areas are expressed in km^2 whilst in the USA acres are commonly used. For volume either m^3 or km^3 are widely used with acre feet used in the USA.

The capacity, volume or storage of a reservoir is usually divided into distinguishable areas. *Dead* or *inactive* storage refers to water in a reservoir that cannot be drained by gravity through a dam's outlet works, spillway or power plant intake and can only be pumped out. Dead storage allows sediments to settle which improves water quality and also creates hydraulic head along with an area for fish during low levels. *Active* or *live* storage is the portion of the reservoir that can be utilised for flood control, power production, navigation and downstream releases. In addition, a reservoir's *flood control capacity* is the amount of water it can regulate during flooding. The *surcharge capacity* is the capacity of the reservoir above the spillway crest that cannot be regulated.

In the United States the water below the normal maximum level of a reservoir is called the *conservation pool*.

In the UK *top water level* describes the reservoir full state whist *fully drawn down* describes the minimum retained volume.

Modelling Reservoir Management

There is a wide variety of software for modelling reservoirs, from the specialist Dam Safety Programme Management Tools (DSPMT) to the relatively simple WAFLEX, to integrated models like the Water Evaluation And Planning system (WEAP) that place reservoir operations in the context of system-wide demands and supplies.

Safety

In many countries large reservoirs are closely regulated to try to prevent or minimise failures of containment.

Whilst much of the effort is directed at the dam and its associated structures as the weakest part of the overall structure, the aim of such controls is to prevent an uncontrolled release of water from the reservoir. Reservoir failures can generate huge increases in flow down a river valley with the potential to wash away towns and villages and cause considerable loss of life such as the devastation following the failure of containment at Llyn Eigiau which killed 17 people.

A notable case of reservoirs being used as an instrument of War involved the British Royal Air Force Dambusters raid on Germany in World War II (codenamed "Operation Chastise"), in which three German reservoir dams were selected to be breached in order to impact on German infrastructure and manufacturing and power capabilities deriving from the Ruhr and Eder rivers. The economic and social impact was derived from the enormous volumes of previously stored water that swept down the valleys wreaking destruction. This raid later became the basis for several films.

Environmental Impact

Whole Life Environmental Impact

All reservoirs will have a monetary cost/benefit assessment made before construction to see if the project is worth proceeding with. However, such analysis can often omit the environmental impacts of dams and the reservoirs that they contain. Some impacts such as the greenhouse gas production associated with concrete manufacture are relatively easy to estimate. Other impact on the natural environment and social and cultural effects can be more difficult to assess and to weigh in the balance but identification and quantification of these issues are now commonly required in major construction projects in the developed world.

Climate Change

Depending upon the circumstances, a reservoir built for hydro-electricity generation can either reduce or increase the net production of greenhouse gases. An increase can occur if plant material in the flooded areas decays in an anaerobic environment releasing (methane and carbon dioxide). This apparently counterintuitive position arises because much carbon is released as methane which is approximately 8 time more potent as a greenhouse gas than carbon dioxide. A study for the National Institute for Research in the Amazon found that Hydroelectric reservoirs release a large pulse of carbon dioxide from above-water decay of trees left standing in the reservoirs, especially during the first decade after closing. This elevates the global warming impact of the dams to levels much higher than would occur by generating the same power from fossil fuels. According to the World Commission on Dams report (Dams And Development), when the reservoir is relatively large and no prior clearing of forest in the flooded area was undertaken, greenhouse gas emissions from the reservoir could be higher than those of a conventional oil-fired thermal generation plant. For instance, In 1990, the impoundment behind the Balbina Dam in Brazil (inaugurated in 1987) had over 20 times the impact on global warming than would generating the same power from fossil fuels, due to the large area flooded per unit of electricity generated.

A decrease can occur if the dam is used in place of traditional power generation, since electricity produced from hydroelectric generation does not give rise to any flue gas emissions from fossil fuel combustion (including sulphur dioxide, nitric oxide and carbon monoxide from coal). The Tucurui dam in Brazil (closed in 1984) had only 0.4 times the impact on global warming than would generating the same power from fossil fuels.

Biology

Dams can produce a block for migrating fish, trapping them in one area, producing food and a habitat for various water-birds. They can also flood various ecosystems on land and may cause extinctions.

Human Impact

Dams can severely reduce the amount of water reaching countries downstream of them, causing water stress between the countries, e.g. the Sudan and Egypt, which damages farming businesses in the downstream countries, and reduces drinking water.

Farms and villages, e.g. Ashopton can be flooded by the creation of reservoirs, ruining many livelihoods. For this very reason, worldwide 80 million people (figure is as of 2009) have had to be forcibly relocated due to dam construction.

Limnology

The limnology of reservoirs has many similarities to that of lakes of equivalent size. There are however significant differences. Many reservoirs experience considerable variations in level producing significant areas that are intermittently underwater or dried out. This greatly limits the productivity or the water margins and limits the number of species able to survive in these conditions.

Upland reservoirs tend to have a much shorter residence time than natural lakes and this can lead to more rapid cycling of nutrients through the water body so that they are more quickly lost to the system. This may be seen as a mismatch between water chemistry and water biology with a tendency for the biological component to be more oligotrophic than the chemistry would suggest.

Conversely, lowland reservoirs drawing water from nutrient rich rivers, may show exaggerated eutrophic characteristics because the residence time in the reservoir is much greater than in the river and the biological systems have a much greater opportunity to utilise the available nutrients. Deep reservoirs with multiple level draw off towers can discharge deep cold water into the downstream river greatly reducing the size of any hypolimnion. This in turn can reduce the concentrations of phosphorus released during any annual mixing event and may therefore reduce productivity.

The Dams in front of reservoirs act as knickpoints-the energy of the water falling from them reduces and deposition is a result below the Dams.

Seismicity

The filling (impounding) of reservoirs has often been attributed to reservoir-triggered seismicity (RTS) as seismic events have occurred near large dams or within their reservoirs in the past. These events may have been triggered by the filling or operation of the reservoir and are on a small scale when compared to the amount of reservoirs worldwide. Of over 100 recorded events, early examples include the 60 m (197 ft) tall Marathon Dam in Greece (1929), the 221 m (725 ft) tall Hoover Dam in the U.S. (1935).

Most events involve large dams and small amounts of seismicity. The only four recorded events above a 6.0-magnitude (M_w) are the 103 m (338 ft) tall Koyna Dam in India which registered a M_w of 6.3 along with the 120 m (394 ft) Kremasta Dam in Greece which registered a 6.3-M_w as well. Following those two, the next largest were the 122 m (400 ft) high Kariba Dam in Zambia at 6.25-M_w and the 105 m (344 ft) Xinfengjiang Dam in China at 6.1-M_w.

Disputes occur over when RTS has occurred due to a lack of hydrogeological knowledge at the time of the event. It is accepted though that the infiltration of water into pores and the weight of the reservoir do contribute to RTS patterns. For RTS to occur, there must be a seismic structure near the dam or its reservoir and the seismic structure must be close to failure. Additionally, water must be able to infiltrate the deep rock stratum as the weight of a 100 m (328 ft) deep reservoir will have little impact when compared the deadweight of rock on a crustal stress field which may be located at a depth of 10 km (6 mi) or more.

Micro Climate

Reservoirs may change the local micro-climate increasing humidity and reducing extremes of temperature. Such effects are claimed by some South Australian winerys as increasing the quality of the wine production.

Surface Runoff

Surface runoff is the water flow that occurs when soil is infiltrated to full capacity and excess water from rain, meltwater, or other sources flows over the land. This is a major component of the water cycle. Runoff that occurs on surfaces before reaching a channel is also called a nonpoint source. If a nonpoint source contains man-made contaminants, the runoff is called nonpoint source pollution. A land area which produces runoff that drains to a common point is called a drainage basin. When runoff flows along the ground, it can pick up soil contaminants including, but not limited to petroleum, pesticides, or fertilizers that become discharge or nonpoint source pollution.

Generation

Snow and glacier melt occur only in areas cold enough for these to form permanently. Typically snowmelt will peak in the spring and glacier melt in the summer, leading to pronounced flow maxima in rivers affected by them. The determining factor of the rate of melting of snow or glaciers is both air temperature and the duration of sunlight. In high mountain regions, streams frequently rise on sunny days and fall on cloudy ones for this reason.

In areas where there is no snow, runoff will come from rainfall. However, not all rainfall will produce runoff because storage from soils can absorb light showers. On the extremely ancient soils of Australia and Southern Africa, proteoid roots with their extremely dense networks of root hairs can absorb so much rainwater as to prevent runoff even when substantial amounts of rain fall. In these regions, even on less infertile cracking clay soils, high amounts of rainfall and potential evaporation are needed to generate any surface runoff, leading to specialised adaptations to extremely variable (usually ephemeral) streams.

Infiltration Excess Overland Flow

This occurs when the rate of rainfall on a surface exceeds the rate at which water can infiltrate the ground, and any depression storage has already been filled. This is called infiltration excess overland flow, Hortonian overland flow (after Robert E. Horton), or unsaturated overland flow. This more commonly occurs in arid and semi-arid regions, where rainfall intensities are high and the soil infiltration capacity is reduced because of surface sealing, or in paved areas. This occurs largely in city areas where pavements prevent water infiltration.

Saturation Excess Overland Flow

When the soil is saturated and the depression storage filled, and rain continues to fall, the rainfall will immediately produce surface runoff. The level of antecedent soil moisture is one factor affecting the time until soil becomes saturated. This runoff is called saturation excess overland flow or saturated overland flow. It is also known as Hewlettian runoff.

Antecedent Soil Moisture

Soil retains a degree of moisture after a rainfall. This residual water moisture affects the soil's infiltration capacity. During the next rainfall event, the infiltration capacity will cause the soil to be saturated at a different rate. The higher the level of antecedent soil moisture, the more quickly the soil becomes saturated. Once the soil is saturated, runoff occurs.

Subsurface Return Flow

After water infiltrates the soil on an up-slope portion of a hill, the water may flow laterally through the soil, and exfiltrate (flow out of the soil) closer to a channel. This is called subsurface return flow or throughflow.

As it flows, the amount of runoff may be reduced in a number of possible ways: a small portion of it may evapotranspire; water may become temporarily stored in microtopographic depressions; and a portion of it may become run-on, which is the infiltration of runoff as it flows overland. Any remaining surface water eventually flows into a receiving water body such as a river, lake, estuary or ocean.

Human Influence

Urbanisation increases surface runoff, by creating more impervious surfaces such as pavement and buildings, that do not allow percolation of the water down through the soil to the aquifer. It is instead forced directly into streams or storm water runoff drains, where erosion and siltation can be major problems, even when flooding is not. Increased runoff reduces groundwater recharge, thus lowering the water table and making droughts worse, especially for farmers and others who depend on the water wells.

When anthropogenic contaminants are dissolved or suspended in runoff, the human impact is expanded to create water pollution. This pollutant load can reach various receiving waters such as streams, rivers, lakes, estuaries and oceans with resultant water chemistry changes to these water systems and their related ecosystems. A 2008 report by the United States National Research Council identified urban stormwater as a leading source of water quality problems in the U.S.

Effects of Surface Runoff

Erosion & Deposition

Surface runoff causes erosion of the Earth's surface; deposition is the depositing of erosion. There are four principal types of erosion: splash erosion, gully erosion, sheet erosion and stream bed erosion. Splash erosion is the result of mechanical collision of raindrops with the soil surface. Dislodged soil particles becoming suspended in the surface runoff and carried into streams and rivers. Gully erosion occurs when the power of runoff is strong enough that it cuts a well defined channel. These channels can be as small as one centimetre wide or as large as several meters. Sheet erosion is the overland transport of runoff without a well defined channel. In the case of gully erosion, large amounts of material can be transported in a small time period. Stream bed erosion is the attrition of stream banks or bottoms by rapidly flowing rivers or creeks. Reduced crop productivity usually results from erosion, and these effects are studied in the field of soil conservation. The soil particles carried in runoff vary in size from about .001 millimetre to 1.0 millimetre in diameter. Larger particles settle over short transport distances, whereas small particles can be carried over long distances suspended in the water column. Erosion of silty soils that contain smaller particles generates turbidity and diminishes light transmission, which disrupts aquatic ecosystems.

Entire sections of countries have been rendered unproductive by erosion. On the high central plateau of Madagascar, approximately ten percent of that country's land area, virtually the entire landscape is devoid of vegetation, with erosive gully furrows typically in excess of 50 meters deep and one kilometre wide. Shifting cultivation is a farming system which sometimes incorporates the slash and burn method in some regions of the world. Erosion cause loss of the fertile top soil and reduces the its fertility and quality of the agricultural produce.

Modern industrial farming is another major cause of erosion. In some areas in the American corn belt, more than 50 percent of the original topsoil has been carried away within the last 100 years.

Environmental Effects

The principal environmental issues associated with runoff are the impacts to surface water, groundwater and soil through transport of water pollutants to these systems. Ultimately these consequences translate into human health risk, ecosystem disturbance and aesthetic impact to water resources. Some of the contaminants that create the greatest impact to surface waters arising from runoff are petroleum substances, herbicides and fertilizers. Quantitative uptake by surface runoff of pesticides and other contaminants has been studied since the 1960s, and early on contact of pesticides with water was known to enhance phytotoxicity. In the case of surface waters, the impacts translate to water pollution, since the streams and rivers have received runoff carrying various chemicals or sediments. When surface waters are used as potable water supplies, they can be compromised regarding health risks and drinking water aesthetics (that is, odour, colour and turbidity effects). Contaminated surface waters risk altering the metabolic processes of the aquatic species that they host; these alterations can lead to death, such as fish kills, or alter the balance of populations present. Other specific impacts are on animal mating, spawning, egg and larvae viability, juvenile survival and plant productivity. Some researches show surface runoff of pesticides, such as DDT, can alter the gender of fish species genetically, which transforms male into female fish.

In the case of groundwater, the main issue is contamination of drinking water, if the aquifer is abstracted for human use. Regarding soil contamination, runoff waters can have two important pathways of concern. Firstly, runoff water can extract soil contaminants and carry them in the form of water pollution to even more sensitive aquatic habitats. Secondly, runoff can deposit contaminants on pristine soils, creating health or ecological consequences.

Flooding

Flooding occurs when a watercourse is unable to convey the quantity of runoff flowing downstream. The frequency with which this occurs is described by a return period. Flooding is a natural process, which maintains ecosystem composition and processes, but it can also be altered by land use changes such as river engineering.

Floods can be both beneficial to societies or cause damage. Agriculture along the Nile floodplain took advantage of the seasonal flooding that deposited nutrients beneficial for crops. However, as the number and susceptibility of settlements increase, flooding increasingly becomes a natural hazard. Adverse impacts span loss of life, property damage, contamination of water supplies, loss of crops, and social dislocation and temporary homelessness. Floods are among the most devastating of natural disasters.

Agricultural Issues

A common context of run-off deals with agriculture. When farmland is tilled and bare soil is revealed, rainwater carries billions of tons of topsoil into waterways each year, causing loss of valuable topsoil and adding sediment to produce turbidity in surface waters.

The other context of agricultural issues involves the transport of agricultural chemicals (nitrates, phosphates, pesticides, herbicides etc.) via surface runoff. This result occurs when chemical use is excessive or poorly timed with respect to high precipitation. The resulting contaminated runoff represents not only a waste of agricultural chemicals, but also an environmental threat to downstream ecosystems. The alternative to conventional farming is organic farming which eliminates chemical usage.

Measurement and Mathematical Modeling

Runoff is analysed by using mathematical models in combination with various water quality sampling methods. Measurements can be made using continuous automated water quality analysis instruments targeted on pollutants such as specific organic or inorganic chemicals, pH, turbidity etc. or targeted on secondary indicators such as dissolved oxygen. Measurements can also be made in batch form by extracting a single water sample and conducting any number of chemical or physical tests on that sample. In the 1950s or earlier hydrology transport models appeared to calculate quantities of runoff, primarily for flood forecasting.

Beginning in the early 1970s computer models were developed to analyse the transport of runoff carrying water pollutants, which considered dissolution rates of various chemicals, infiltration into soils and ultimate pollutant load delivered to receiving waters. One of the earliest models addressing chemical dissolution in runoff and resulting transport was developed in the early 1970s under contract to the United States Environmental Protection Agency (EPA). This computer model formed the basis of much of thc mitigation study that led to strategies for land use and chemical handling controls.

Other computer models have been developed (such as the DSSAM Model) that allow surface runoff to be tracked through a river course as reactive water pollutants. In this case the surface runoff may be considered to be a line source of water pollution to the receiving waters.

Spring (Hydrosphere)

A spring—also known as a rising or resurgence—is a component of the hydrosphere. Specifically, it is any natural situation where water flows to the surface of the earth from underground. Thus, a spring is a site where the aquifer surface meets the ground surface.

Formation

The forcing of the spring to the surface can be the result of a confined aquifer in which the recharge area of the spring water table rests at a higher elevation than that of the outlet. Spring water forced to the surface by elevated sources are artesian wells. This is possible even if the outlet is in the form of a 300-foot-deep (91 m) cave.

In this case the cave is used like a hose by the higher elevated recharge area of groundwater to exit through the lower elevation opening. Non-artesian springs may simply flow from a higher elevation through the earth to a lower elevation and exit in the form of a spring, using the ground like a drainage pipe. Still other springs are the result of pressure from an underground source in the earth, in the form of volcanic activity. The result can be water at elevated temperature such as a hot spring.

The action of the groundwater continually dissolves permeable bedrock such as limestone and dolomite creating vast cave systems.

Types of Spring Outlets

- Seepage or filtration spring. The term seep refers to springs with small flow rates in which the source water has filtered into permeable earth.
- Fracture springs, discharge from faults, joints, or fissures in the earth, in which springs have followed a natural course of voids or weaknesses in the bedrock.
- Tubular springs are essentially water dissolved and created underground channels, basically cave systems.

Flow

Spring discharge, or resurgence, is determined by the spring's recharge basin. Factors that affect the recharge include the size of the area in which groundwater is captured, the amount of precipitation, the size of capture points, and the size of the spring outlet. Water may leak into the underground system from many sources including permeable earth, sinkholes, and losing streams.

In some cases entire creeks seemingly disappear as the water sinks into the ground via the stream bed. Grand Gulf State Park in Missouri is an example of an entire creek vanishing into the groundwater system. The water emerges nine miles (14 km) away, forming some of the discharge of Mammoth Spring in Arkansas. Human activity may also affect a spring's discharge-withdraw of groundwater reduces the water pressure in an aquifer, decreasing the volume of flow.

Classification

Springs are often classified by the volume of the water they discharge. The largest springs are called "first-magnitude," defined as springs that discharge water at a rate of at least 2800 litres or 100 cubic feet (2.8 m^3) of water per second.

Magnitude	*Flow (ft^3/s, gal/min, pint/min)*	*Flow (L/s)*
1st magnitude	> 100 ft^3/s	2800 L/s
2nd magnitude	10 to 100 ft^3/s	280 to 2800 L/s
3rd magnitude	1 to 10 ft^3/s	28 to 280 L/s
4th magnitude	100 US gal/min to 1 ft^3/s	
	(448 US gal/min)	6.3 to 28 L/s
5th magnitude	10 to 100 gal/min	0.63 to 6.3 L/s
6th magnitude	1 to 10 gal/min	63 to 630 mL/s
7th magnitude	1 pint to 1 gal/min	8 to 63 mL/s
8th magnitude	Less than 1 pint/min	8 mL/s
0 magnitude	no flow (sites of past/historic flow)	

Some locations contain many first-magnitude springs, such as Florida where there are at least 27 known to be that size, the Missouri and Arkansas Ozarks contain (10 known of first-magnitude), and 11 more in the Thousand Springs area along the Snake River in Idaho. The scale for spring flow is as follows:

Water Content

Minerals become dissolved in the water as it moves through the underground rocks. This may give the water flavour and even carbon dioxide bubbles, depending on the nature of the geology through which it passes. This is why spring water is often bottled and sold as mineral water, although the term is often the subject of deceptive advertising. Springs that contain significant amounts of minerals are sometimes called 'mineral springs'. Springs that contain large amounts of dissolved sodium salts, mostly sodium carbonate, are called 'soda springs'. Many resorts have developed around mineral springs and are known as spa towns.

Water from springs are usually clear. However some springs may be coloured by the minerals that are dissolved in the water. Iron and tannins often give spring water an orange colour.

A stream carrying the outflow of a spring to a nearby primary stream is called a spring branch or run. Groundwater tends to maintain a relatively long-term average temperature of its aquifer; so flow from a spring may be cooler than a summer day, but remain unfrozen in the winter. The cool water of a spring and its branch may harbour species such as certain trout that are otherwise ill-suited to a warmer local climate.

Historic Uses

Springs have been used for a variety of human needs including drinking water, powering of mills, and navigation, and more recently some have been used for electricity generation.

Sacred Springs

A sacred spring, or holy well, is a small body of water emerging from underground and revered either in a pagan or a Christian context, often both. The lore and mythology of ancient Greece was replete with sacred and storied springs—notably, the Corycian, Pierian and Castalian. In medieval Europe, holy wells were frequently pagan sacred sites that later became Christianised. The term "holy well" is commonly employed to refer to any water source of limited size (i.e. not a lake or river, but including pools and natural springs and seeps), which has some significance in the folklore of the local area. This can take the form of a particular name, an associated legend, the attribution of healing qualities to the water through the numinous presence of its guardian spirit or Christian saint, or a ceremony or ritual centred on the well site. In Christian legend, the spring water is often said to have been made to flow by the action of a saint, a familiar theme especially in the hagiography of Celtic saints.

Modern Uses

Present-day uses include recreational activities, such as trout-fishing, swimming, and floating. The water is also used for livestock and fish hatcheries.

Iceberg

An iceberg is a large piece of ice from freshwater that has broken off from a snow-formed glacier or ice shelf and is floating in open water. It may subsequently become frozen into pack ice. Alternatively, it may come to rest on the seabed in shallower water, causing ice scour (also known as ice gouging) or becoming an *ice island.*

Etymology

The word "iceberg" is a partial loan translation from Dutch *ijsberg*, literally meaning *ice mountain*, cognate to Danish *Isbjerg*, German *Eisberg*, Low Saxon *Iesbarg*, Swedish *Isberg* and Norwegian *Isfjell.*

Overview

Because the density of pure ice is about 920 kg/m^3, and that of sea water about 1025 kg/m^3, typically only one-ninth of the volume of an iceberg is above water. The shape of the underwater portion can be difficult to judge by looking at the portion above the surface. This has led to the expression "tip of the iceberg", for a problem or difficulty that is only a small manifestation of a larger problem.

Icebergs generally range from 1 to 75 metres (3.3 to 246 ft) above sea level and weigh 100,000 to 200,000 metric tons (110,000 to 220,000 short tons). The tallest known iceberg in the North Atlantic was 168 metres (551 ft) above sea level, reported by the USCG icebreaker *East Wind* in 1958, making it the height of a 55-storey building. These icebergs originate from the glaciers of western Greenland, and may have an interior temperature of -15 to -20°C (5 to -4 °F).

Though usually confined by winds and currents to move close to the coast, the largest icebergs recorded have been calved, or broken off, from the Ross Ice Shelf of Antarctica. Iceberg B-15, photographed by satellite in 2000, measured 295 by 37 kilometres (183 by 23 mi), with a surface area of 11,000 square kilometres (4,200 sq mi). The mass was estimated around three billion tonnes. The largest iceberg on record was an Antarctic tabular iceberg of over 31,000 square kilometres (12,000 sq mi) [335 by 97 kilometres (208 by 60 mi)] sighted

150 miles (240 km) west of Scott Island, in the South Pacific Ocean, by the USS *Glacier* on November 12, 1956. This iceberg was larger than Belgium.

When an iceberg melts, it makes a fizsing sound called "Bergie Seltzer". This sound is made when compressed air bubbles trapped in the iceberg pop. The bubbles come from air trapped in snow layers that later became glacial ice.

Classification

Size: Names for various sizes of iceberg are not universal, but usually follow a similar pattern. The size classification in the table below is used by the International Ice Patrol:

Size Category	*Height*	*Length*
Growler	Less than 1 metre (3.3 ft)	Less than 5 metres (16 ft)
Bergy Bit	1–5 metres (3.3–16 ft)	5–15 metres (16–49 ft)
Small	5–15 metres (16–49 ft)	15–60 metres (49–200 ft)
Medium	15–45 metres (49–148 ft)	60–120 metres (200–390 ft)
Large	45–75 metres (148–246 ft)	120–200 metres (390–660 ft)
Very Large	Over 75 metres (246 ft)	Over 200 metres (660 ft)

Recent Large Icebergs

- Iceberg B-15 11,000 km^2 (4,200 sq mi), 2000
- Iceberg A-38, about 6,900 km^2 (2,700 sq mi), 1998
- Iceberg B-15A, 3,100 km^2 (1,200 sq mi), broke off 2003
- Iceberg C-19, 5,500 km^2 (2,100 sq mi), 2002
- Iceberg B-9, 5,390 km^2 (2,080 sq mi), 1987
- Iceberg D-16, 310 km^2 (120 sq mi), 2006
- Ice sheet, 260 km^2 (100 sq mi), broken off of Petermann Glacier in northern Greenland on Aug 5, 2010, considered to be largest Arctic iceberg since 1962. About a month later, this iceberg split into two pieces upon crashing into Joe Island in the Nares Strait next to Greenland. In June 2011, large fragments of the Petermann Ice Islands were observed off the Labrador coast.
- Iceberg B-17B 140 km^2 (54 sq mi), 1999, shipping alert issued December 2009.

Shape

In addition to size classification, Icebergs can also be classified on the basis of their shape. The two basic types of iceberg forms are *tabular* and *non-tabular*. Tabular icebergs have steep sides and a flat top, much like a plateau, with a length-to-height ratio of more than 5:1. This type of iceberg can be quite large, as in the case of Pobeda Ice Island. Antarctic icebergs formed by breaking off from an ice shelf, such as the Ross Ice Shelf or Filchner-Ronne Ice Shelf, are typically tabular. The largest icebergs in the world are formed this way.

Non-tabular icebergs have different shapes, and include:

- *Dome*: An iceberg with a rounded top.
- *Pinnacle*: An iceberg with one or more spires.
- *Wedge*: An iceberg with a steep edge on one side and a slope on the opposite side.
- *Dry-Dock*: An iceberg that has eroded to form a slot or channel.
- *Blocky*: An iceberg with steep, vertical sides and a flat top. It differs from tabular icebergs in that its shape is more like a block than a flat sheet.

Monitoring

Icebergs are monitored worldwide by the U.S. National Ice Centre (NIC), established in 1995, which produces analyses and forecasts of Arctic, Antarctic, Great Lakes and Chesapeake Bay ice conditions. More than 95 per cent of the data used in its sea ice analyses are derived from the remote sensors on polar-orbiting satellites that survey these remote regions of the Earth.

The NIC is the only organisation that names and tracks all Antarctic Icebergs. It assigns each iceberg larger than 10 nautical miles (19 km) along at least one axis a name composed of a letter indicating its point of origin and a running number. The letters used are as follows:

- A – longitude 0° to 90° W (Bellingshausen Sea, Weddell Sea)
- B – longitude 90° W to 180° (Amundsen Sea, Eastern Ross Sea)
- C – longitude 90° E to 180° (Western Ross Sea, Wilkes Land)
- D – longitude 0° to 90° E (Amery Ice Shelf, Eastern Weddell Sea).

Iceberg B15 calved from the Ross Ice Shelf in 2000 and initially had an area of 11,000 square kilometres (4,200 sq mi). It broke apart in November 2002. The largest remaining piece of it, Iceberg B-15A, with an area of 3,000 square kilometres (1,200 sq mi), was still the largest iceberg on Earth until it ran aground and split into several pieces October 27, 2005. It has been determined that the cause of the break-up was an ocean swell generated by an Alaskan storm 6 days earlier and 13,500 kilometres (8,400 mi) away.

History

In the 20th century, several scientific bodies were established to study and monitor the icebergs. The International Ice Patrol, formed in 1914 in response to the *Titanic* disaster, monitors iceberg dangers near the Grand Banks of Newfoundland and provides the "limits of all known ice" in that vicinity to the maritime community.

Technology History

Before April 1912 there was no system in place to track icebergs to guard ships against collisions. The sinking of the RMS *Titanic*, which caused the deaths of 1,523 of its 2,228 passengers, created the demand for a system to observe icebergs. For the remainder of the ice season of that year, the United States Navy patrolled the waters and monitored ice flow. In November 1913, the International Conference on the Safety of Life at Sea met in London to devise a more permanent system of observing icebergs. Within three months the participating maritime nations had formed the International Ice Patrol (IIP). The goal of the IIP was to collect data on meteorology and oceanography in order to measure currents, ice-flow, ocean temperature, and salinity levels. They published their first records in 1921, which allowed for a year-by-year comparison of iceberg movement.

New technologies monitor icebergs. Aerial surveillance of the seas in the early 1930s allowed for the development of charter systems that could accurately detail the ocean currents and iceberg locations. In 1945, experiments tested the effectiveness of radar in detecting icebergs. A decade later, oceanographic monitoring outposts were established for the purpose of collecting data; these outposts continue to serve in environmental study. A computer was first installed on a ship for the purpose of oceanographic monitoring in 1964, which allowed for a faster evaluation of data. By the 1970s, icebreaking ships were equipped with automatic transmissions of satellite photographs of ice in Antarctica. Systems for optical satellites had been developed, but were still limited by weather conditions.

In the 1980s, drifting buoys were used in Antarctic waters for oceanographic and climate research. They are equipped with sensors that measure ocean temperature and currents. Side-Looking Airborne Radar (SLAR) made it possible to acquire images regardless of weather conditions. On November 4, 1995, Canada launched RADARSAT-1. Developed by the Canadian Space Agency, it provides images of Earth for both scientific and commercial purposes. This system was the first to use Synthetic Aperture Radar (SAR), which sends microwave energy to the ocean surface and records the reflections to track icebergs. The European Space Agency launched ENVISAT on March 1, 2002, an environmental satellite which uses Advanced Synthetic Aperture Radar (ASAR). This can detect changes in surface height accurately. The Canadian Space Agency launched RADARSAT-2 in December 2007, which uses SAR and multipolarisation modes and follows the same orbit path as RADARSAT-1.

5

Natural Biological Treatment Systems

Natural low-rate biological treatment systems are available for the treatment of organic wastewaters such as municipal sewage and tend to be lower in cost and less sophisticated in operation and maintenance. Although such processes tend to be land intensive by comparison with the conventional high-rate biological processes already described, they are often more effective in removing pathogens and do so reliably and continuously if properly designed and not overloaded. Among the natural biological treatment systems available, stabilisation ponds and land treatment have been used widely around the world and a considerable record of experience and design practice has been documented. The nutrient film technique is a fairly recent development of the hydroponic plant growth system with application in the treatment and use of wastewater.

Wastewater Stabilisation Ponds

A recent World Bank Report (Shuval *et al.* 1986) came out strongly in favour of stabilisation ponds as the most suitable wastewater treatment system for effluent use in agriculture. Stabilisation ponds are the preferred wastewater treatment process in developing countries, where land is often available at reasonable opportunity cost and skilled labour is in short supply.

Wastewater stabilisation pond systems are designed to achieve different forms of treatment in up to three stages in series, depending on the organic strength of the input waste and the effluent quality objectives. For ease of maintenance and flexibility of operation, at least two trains of ponds in parallel are incorporated in any design. Strong wastewaters, with BOD_5 concentration in excess of about 300 mg/l, will frequently be introduced into first-stage anaerobic ponds, which achieve a high volumetric rate of removal. Weaker wastes or, where anaerobic ponds are environmentally unacceptable, even stronger wastes (say up to 1000 mg/l BOD_5) may be discharged directly into primary facultative ponds.

Effluent from first-stage anaerobic ponds will overflow into secondary facultative ponds which comprise the second-stage of biological treatment. Following primary or secondary facultative ponds, if further pathogen reduction is necessary maturation ponds will be introduced to provide tertiary treatment.

***Anaerobic Ponds*:** Anaerobic ponds are very cost effective for the removal of BOD, when it is present in high concentration. Normally, a single, anaerobic pond in each treatment train is sufficient if the strength of the influent wastewater, L_i is less than 1000 mg/l BOD_5. For high strength industrial wastes, up to three anaerobic ponds in series might be justifiable but the retention time t_{an}, in any of these ponds should not be less than 1 day (McGarry and Pescod, 1970).

Anaerobic conditions in first-stage stabilisation ponds are created by maintaining a high volumetric organic loading, certainly greater than 100g BOD_5/m^3 d. Volumetric loading, λ_v, is given by:

$$\lambda = \frac{L_i Q}{V}$$

where:

L_i = Influent BOD_5, mg/l,

Q = Influent flow rate, m^3/d, and

V = Pond volume, m^3

or, since V/Q = t_{an}, the retention time:

$$\lambda_v = \frac{L_i}{t_{an}}$$

Very high loadings, up to 1000g BOD_5/m^3d, achieve efficient utilisation of anaerobic pond volume but, with wastewater containing sulphate concentrations in excess of 100 mg/l, the production of H_2S is likely to cause odour problems. In the case of typical municipal sewage, it is generally accepted that a maximum anaerobic pond loading of 400g BOD_5/m^3d will prevent odour nuisance (Meiring et al. 1968).

Anaerobic ponds normally have a depth between 2m and 5m and function as open septic tanks with gas release to the atmosphere. The biochemical reactions which take place in anaerobic ponds are the same as those occurring in anaerobic digesters, with a first phase of acidogenesis and a second slower-rate of methanogenesis. Ambient temperatures in hot-climate countries are conducive to these anaerobic reactions and expected BOD_5 removals for different retention times in treating sewage have been given by Mara (1976).

More recently, Gambrill et al. (1986) have suggested conservative removals of BOD_5 in anaerobic ponds as 40 per cent below 10°C, at a design loading, l_v, of 100 g/m³d, and 60 per cent above 20°C, at a design loading of 300 g/m³d, with linear interpolation for operating temperature between 10 and 20°C. Higher removal rates are possible with industrial wastes, particularly those containing significant quantities of organic settleable solids. Of course, other environmental conditions in the ponds, particularly pH, must be suitable for the anaerobic microorganisms bringing about the breakdown of BOD.

Table: Bod Removals in Anaerobic Ponds Loaded at 250 g BOD_5/m^3d

Retention t_{an} days	***BOD_5 removal %***
1	50
2.5	60
5	70

Source: Mara (1976)

In certain instances, anaerobic ponds become covered with a thick scum layer, which is thought to be beneficial but not essential, and may give rise to increased fly breeding. Solids in the raw wastewater, as well as biomass produced, will settle out in first-stage anaerobic ponds and it is common to remove sludge when it has reached half depth in the pond. This usually occurs after two years of operation at design flow in the case of municipal sewage treatment.

Facultative Ponds: The effluent from anaerobic ponds will require some form of aerobic treatment before discharge or use and facultative ponds will often be more appropriate than conventional forms of secondary biological treatment for application in developing countries. Primary facultative ponds will be designed for the treatment of weaker wastes and in sensitive locations where anaerobic pond odours would be unacceptable. Solids in the influent to a facultative pond and excess biomass produced in the pond will settle out forming a sludge layer at the bottom. The benthic layer will be anaerobic and, as a result of anaerobic breakdown of organics, will release soluble organic products to the water column above.

Organic matter dissolved or suspended in the water column will be metabolised by heterotrophic bacteria, with the uptake of oxygen, as in convential aerobic biological wastewater treatment processes. However, unlike in convential processes, the dissolved oxygen utilised by the bacteria in facultative ponds is replaced through photosynthetic oxygen production by microalgae, rather than by aeration equipment. Especially intreating municipal sewage in hot climates, the environment in facultative ponds is ideal for the proliferation of microalgae. High temperature and ample sunlight create conditions which encourage algae to utilise the carbon dioxide (CO_2) released by bacteria in breaking down the organic components of the wastewater and take up nutrients (mainly nitrogen and phosphorus) contained in the wastewater. This symbiotic relationship contributes to the overall removal of BOD in facultative ponds.

To maintain the balance necessary to allow this symbiosis to persist, the organic loading on a facultative pond must be strictly limited. Even under satisfactory operating conditions, the dissolved oxygen concentration (DO) in a facultative pond will vary diurnally as well as over the depth. Maximum DO will occur at the surface of the pond and will usually reach supersaturation in tropical regions at the time of maximum radiation intensity. From that time until sunrise, DO will decline and may well disappear completely for a short period. For a typical facultative pond depth, D_f, of 1.5m the water column will be predominantly aerobic at the time of peak radiation and predominantly anaerobic at sunrise. The pH of the pond contents will also vary diurnally as algae utilise CO_2 throughout daylight hours and respire, along with bacteria and other organisms, releasing CO_2 during the night. Wind is considered important to the satisfactory operation of facultative ponds by mixing the contents and helping to prevent short-circuiting. Intimate mixing of organic substrate and the degrading organisms is important in any biological reactor but in facultative ponds wind mixing is considered essential to prevent thermal stratification causing anaerobiosis and failure. Facultative ponds should be orientated with the longest dimension in the direction of the prevailing wind.

Although completely-mixed reactor theory with the assumption of first-order kinetics for BOD removal can be adopted for facultative pond design (Marais and Shaw, 1961), such a fundamental approach is rarely adopted in practice. Instead, an empirical procedure based on operational experience is more common. The most widely adopted design method currently being applied wherever local experience is limited is that introduced by McGarry and Pescod (1970). A regression analysis of operating data on ponds around the world relating maximum surface organic loading, in lb/acre d, to the mean ambient air temperature, in °F, of the coldest month resulted in the following equation (now converted to metric units):

$$\lambda_{s(\max)} = 60.3(1.099)^T$$

where:

λ_s = surface or arael organic loading, kg BOD_5/ha.d

T = mean ambient air temperature of coldest month,°C

Subsequently, Arthur (1983) modified this formula and suggested that best agreement with available operating data, including a factor of safety of about 1.5, is represented by the relationship:

$$\lambda_s = 20T - 60$$

This surface (or areal) BOD_5 loading can be translated into a mid-depth facultative pond area requirement (A_f in m^2) using the formula:

$$A_f = \frac{10 L_i Q}{\lambda_s}$$

Thus:

$$A_f = \frac{L_i Q}{2T - 6}$$

and the mean hydraulic retention time in the facultative pond (t_f in days) is given by:

$$t_f = \frac{A_f D_f}{Q}$$

The removal of BOD_5 in facultative ponds (l_r in kg/ha d) is related to BOD_5 loading and usually averages 70-80 per cent of l_s. Retention time in a properly designed facultative pond will normally be 20-40 days and, with a depth of about 1.5m, the area required will be signficantly greater than for an anaerobic pond. The effluent from a facultative pond treating municipal sewage in the tropics will normally have a BOD_5 between 50 and 70 mg/l as a result of the suspended algae. On discharge to a surface water, this effluent will not cause problems downstream if the dilution is of the order of 8:1 and any live algae in the effluent might well be beneficial as a result of photo synthetic oxygen production during daylight hours.

Efficiently operating facultative ponds treating wastewater will contain a mixed population of flora but flagellate algal genera such as Chlamydomonas, Euglena, Phacus and Pyrobotrys will predominate. Non-motile forms such as Chlorella, Scenedesmus and various diatom species will be present in low concentrations unless the pond is underloaded. Algal stratification often occurs in facultative ponds, particularly in the absence of wind-induced mixing, as motile forms respond to changes in light intensity and move in a band up and down the water column. The relative numbers of different genera and their dominance in a facultative pond vary from season to season throughout the year but species diversity generally decreases with increase in loading. Sometimes, mobile purple sulphur bacteria appear when facultative ponds are overloaded and sulphide concentration increases, with the danger of odour production. High ammonia concentrations also bring on the same problem and are toxic to algae, especially above pH 8.0.

Maintenance of properly designed facultative ponds will be limited to the removal of scum mats, which tend to accumulate in downwind corners, and the cutting of grass on embankments. To ensure efficient operation, facultative ponds should be regularly monitored but, even where this is not possible, they have the reputation of being relatively trouble-free.

***Maturation Ponds*:** The effluent from facultative ponds treating municipal sewage or equivalent input wastewater will normally contain at least 50 mg/l BOD_5 and if an effluent with lower BOD_5 concentration is required it will be necessary to use maturation ponds. For sewage treatment, two maturation ponds in series, each with a retention time of 7 days, have been found necessary to produce a final effluent with $BOD_5 < 25$ mg/l when the facultative pond effluent had a $BOD_5 < 75$ mg/l.

A more important function of maturation ponds, however, is the removal of excreted pathogens to achieve an effluent quality which is suitable for its downstream reuse. Although the longer retention in anaerobic and facultative pond systems will make them more efficient than conventional wastewater treatment processes in removing pathogens, the effluent from a facultative pond treating municipal sewage will generally require further treatment in maturation ponds to reach effluent standards imposed for reuse in unrestricted irrigation. Faecal coliform bacteria are commonly used as indicators of excreted pathogens and maturation ponds can be designed to achieve a given reduction of faecal coliforms (FC). Protozoan cysts and helminth ova are removed by sedimentation in stabilisation ponds and a series of ponds with overall retention of 20 days or more will produce an effluent totally free of cysts and ova (Feachem *et al.* 1983).

Reduction of faecal coliform bacteria in any stabilisation pond (anaerobic, facultative and maturation) is generally taken as following first-order kinetics:

$$N_e = \frac{N_i}{1 + K_b t}$$

where:

N_e = Number of faecal coliforms/100 ml of effluent

N_i = Number of faecal coliforms/100 ml of influent

K_b = First-order rate constant for FC removal, d^{-1}

t = Retention time in any pond, d

For n ponds in series, becomes:

$$N_e = \frac{N_i}{(1 + k_b t_{an})\left(1 + k_b t_f\right)\left(1 + k_b t_{m_1}\right) \cdots \left(1 + k_b t_{m_0}\right)}$$

where:

$t_{m\,n}$ = Retention time in the nth maturation pond.

The value of KB is extremely sensitive to temperature and was shown by Marais (1974) to be given by:

where:

$$K_{b(T)} = 2.6(1.19)^{T-20}$$

$K_{b(T)}$ = value of KB at T°C

A suitable design value of N_i in the case of municipal sewage treatment is 1 x 10^8 faecal coliforms/100ml, which is slightly higher than average practical levels.

The value of N_e should be obtained by substituting the appropriate levels of variables assuming a retention time of 7 days in each of two maturation lagoons (for sewage). If the calculated value of N_e does not meet the reuse effluent standard, the number of maturation ponds should be increased, say to three or more each with retention time 5 days, and N_e recalculated. A more systematic approach is now available whereby the optimum design for maturation ponds can be obtained using a simple computer programme (Gambrill *et al.* 1986).

Polprasert *et al.* (1983) have published an approach to the assessment of bacterial die-off which attempts to take into account the complex physical characteristics of ponds and biochemical reactions taking place in them. A multiple-regression equation involving parameters such as retention time, organic loading, algal concentration and ultra-violet light exposure has been suggested. The Wehner and Wilhelm (1956) non-ideal flow equation, including the pond dispersion number, was adopted to predict bacterial survival, in preference to the first order rate equation (Eq 9 and 10).

Maturation ponds will be aerobic throughout the water column during daylight hours and the pH will rise above 9.0. The algal population of many species of non-flagellate unicellular and colonial forms will be distributed over the full depth of a maturation pond. Large numbers of filamentous algae, particularly blue-greens, will emerge under very low BOD loading conditions. Very low concentrations of algae in a maturation pond will indicate excessive algal predation by zooplankton, such as Daphnia sp, and this will have a deleterious effect on pathogen die-off, which is linked to algal activity.

Saqqar (1988), in his analysis of the performance of the Al Samra stabilisation ponds in Amman, Jordan, has shown that the coliform and faecal coliform die-off coefficients varied with retention time, water temperature, organic loading, total BOD_5 concentration, pH and pond depth. Total coliform die-off was less than the rate of faecal coliform die-off, except during the cold season. For the series of ten ponds, including at least the first five totally anaerobic, the faecal coliform die-off coefficient, k, for the temperature range

12 - 15°C increased through the pond sequence from 0.11 per day in the first anaerobic pond to 0.68 per day in the final two ponds, which operated as facultative ponds.

Overland Treatment of Wastewater

Apart from the use of effluent for irrigation of crops, termed 'slow rate' land treatment in the US Environmental Protection Agency's Process Design Manual for Land Treatment of Municipal Wastewaters (EPA 1977), and 'rapid infiltration' or 'infiltration percolation' of effluent discussed as soil-aquifer treatment in a section of this document, the EPA manual deals with 'overland flow' as a wastewater treatment method. In overland flow treatment, effluent is distributed over gently sloping grassland on fairly impermeable soils. Ideally, the wastewater moves evenly down the slope to collecting ditches at the bottom edge of the area and water-tolerant grasses are an essential component of the system.

This form of land treatment requires alternating applications of effluent (usually treated) and resting of the land, to allow soil reaction and grass cutting. The total area utilised is normally broken up into small plots to allow this form of intermittent operation and yet achieve continuous treatment of the flow of wastewater. Although this type of land treatment has been widely adopted in Australia, New Zealand and the UK for tertiary upgrading of secondary effluents, it has been used for the treatment of primary effluent in Werribee, Australia and is being considered for the treatment of raw sewage in Karachi, Pakistan.

Table: *Site Characteristics and Design Features for Overland Flow Treatment of Wastewater*

Grade	*Finished slope 2-8%*
Field area required (ha)	6.55-44
Soil permeability	Slow (clays, silts and soils with impermeable barriers)
Annual application rate (m)	3-20
Typical weekly application rate (cm)	6-40

Source: EPA (1977)

It was pointed out that steeper land slopes might be feasible at reduced hydraulic loadings. The ranges given for field area required and application rates cover the wastewater quality from raw sewage to secondary effluent, with higher application rates and lower land area requirements being associated with higher levels of preapplication treatment. Although soil permeability is not critical with this form of land treatment, the impact on groundwater should not be overlooked in the case of highly permeable soils.

The application rate for wastewaters will depend principally on the type of soil, the quality of wastewater effluent and the physical and biochemical activity in the near-surface environment. Rational design procedures, based on the kinetics of BOD removal, have been developed for overland flow systems by Middlebrooks *et al.* (1982). Slope lengths from 30 - 60 m are common in the US for overland flow systems.

The cover crop is an important component of the overland flow system since it prevents soil erosion, provides nutrient uptake and serves as a fixed-film medium for biological treatment. Crops best suited to overland flow treatment are grasses with a long growing season, high moisture tolerance and extensive root formation. Reed canary grass has a very high nutrient uptake capacity and yields a good quality hay; other suitable grasses include rye grass and tall fescue.

Suspended and colloidal organic materials in the wastewater are removed by sedimentation and filtration through surface grass and organic layers. Removal of total nitrogen and ammonia is inversely related to application rate, slope length and soil temperature. Phosphorus and trace elements removal is by sorption on soil clay colloids and precipitation as insoluble complexes of calcium, iron and aluminium. Overland flow systems also remove pathogens from sewage effluent at levels comparable with conventional secondary treatment systems, without chlorination. A monitoring programme should always be incorporated into the design of overland flow projects both for wastewater and effluent quality and for application rates.

Macrophyte Treatment

Maturation ponds which incorporate floating, submerged or emergent aquatic plant species are termed *macrophyte ponds* and these have been used in recent years for upgrading effluents from stabilisation ponds. Macrophytes take up large amounts of inorganic nutrients (especially N and P) and heavy metals (such as Cd, Cu, Hg and Zn) as a consequnce of the growth requirements and decrease the concentration of algal cells through light shading by the leaf canopy and, possibly, adherence to gelatinous biomass which grows on the roots.

Floating macrophyte systems utilising water hyacinth and receiving primary sewage effluent in Florida have achieved secondary treatment effluent quality with a 6 day hydraulic retention time, water depth of 60 cm and hydraulic loading 1860 m^3/ha d (Reddy and Debusk 1987). The same authors suggested that similar results had also been observed for artificial wetlands using emergent macrophytes. In Europe, the land area considered to be necessary for treatment of preliminary-treated sewage is estimated at 2-5 m^2 per population equivalent to achieve a secondary effluent quality (Cooper et al. 1988).

***Floating Aquatic Macrophyte Systems*:** Floating macrophyte species, with their large root systems, are very efficient at nutrient stripping. Although several genera have been used in pilot schemes, including *Salvinia, Spirodella, Lemna* and *Eichornia* (O'Brien 1981), *Eichornia crassipes* (water hyacinth) has been studied in much greater detail. In tropical regions, water hyacinth doubles in mass about every 6 days and a macrophyte pond can produce more than 250 kg/ha d^{-1} (dry weight). Nitrogen and phosphorus reductions up to 80 per cent and 50 per cent have been achieved. In Tamil Nadu, India, studies have indicated that the coontail, *Ceratophyllum demersum,* a submerged macrophyte, is very efficient at removing ammonia (97%) and phosphorus (96%) from raw sewage and also removes 95 per cent of the BOD_5. It has a lower growth rate than *Eichornia crassipes,* which allows less frequent harvesting.

In such macrophyte pond systems, apart from any physical removal processes which might occur (especially sedimentation) the aquatic vascular plants serve as living substrates for microbial activity, which removes BOD and nitrogen, and achieves reductions in phosphorus, heavy metals and some organics through plant uptake. The basic function of the macrophytes in the latter mechanism is to assimilate, concentrate and store contaminants on a short-term basis. Subsequent harvest of the plant biomass results in permanent removal of stored contaminents from the pond treatment system.

The nutrient assimilation capacity of aquatic macrophytes is directly related to growth rate, standing crop and tissue composition. The potential rate of pollutant storage by an aquatic plant is limited by the growth rate and standing crop of biomass per unit area. Water hyacinth, for example, was found to reach a standing crop level of 30 tonnes (dry weight)/ha in Florida, resulting in a maximum storage of 900 kg N/ha and 180 kg P/ha (Reddy and De Busk 1987).

Fly and mosquito breeding is a problem in floating macrophyte ponds but this can be partially alleviated by introducing larvae-eating fish species such as *Gambusia* and *Peocelia* into the ponds. It should be recognised that pathogen die-off is poor in macrophyte ponds as a result of light shading and the lower dissolved oxygen and pH compared with algal maturation ponds. In their favour, macrophyte ponds can serve a useful purpose in stripping pond effluents of nutrients and algae and at the same time produce a harvestable biomass. Floating macrophytes are fairly easily collected by floating harvesters. The harvested plants might be fed to cattle, used as a green manure in agriculture, composted aerobically to produce a fertilizer and soil conditioner, or can be converted into biogas in an anaerobic digester, in which case the residual sludge can then be applied as a fertilizer and soil conditioner (UN Economic and Social Commission for Asia and the Pacific 1981). Maximum removal by water hyacinth was 5850 kg N/ha year, compared with 1200 kg N/ha year by duckweed.

Emergent Macrophyte Treatment Systems: In recent years, natural and artificial wetlands and marshes have been used to treat raw sewage and partially-treated effluents. Natural wetlands are usually unmanaged, whereas artificial systems are specially designed to maximise performance by providing the optimum conditions for emergent macrophyte growth. The key features of such reed bed treatment systems are:

- Rhizomes of the reeds grow vertically and horizontally in the soil or gravel bed, opening up 'hydraulic pathways'.
- Wastewater BOD and nitrogen are removed by bacterial activity; aerobic treatment takes place in the rhizosphere, with anoxic and anaerobic treatment taking place in the surrounding soil.
- Oxygen passes from the atmosphere to the rhizosphere via the leaves and stems of the reeds through the hollow rhizomes and out through the roots.
- Suspended solids in the sewage are aerobically composted in the above-ground layer of vegetation formed from dead leaves and stems.
- Nutrients and heavy metals are removed by plant uptake.

The growth rate and pollutant assimilative capacity of emergent macrophytes such as *Phragmites communis and Scirpus lacstris* are limited by the culture system, wastewater loading rate, plant density, climate and management factors. High tissue N concentrations have been found in plants cultured in nutrient enriched (wastewater) systems and in plants analyzed in the early stages of growth. Maximum storage of nutrients by emergent macrophytes was found to be in the range 200-1560 kg N/ha and 40-375 kg P/ha in Florida (Reddy and DeBusk 1987). More than 50 percent of the nutrients were stored in below-ground portions of the plants, tissues difficult to harvest to achieve effective nutrient removal. However, because emergent macrophytes have more supportive tissue than floating macrophytes, they might have greater potential for storing the nutrients over a longer period. Consequently, frequent harvesting might not be so necessary to achieve maximum nutrient removal although harvesting above-ground biomass once a year should improve overall nutrient removal efficiency.

Nutrient Film Technique

The nutrient film technique (NFT) is a modification of the hydroponic plant growth system in which plants are grown directly on an impermeable surface to which a thin film of wastewater is continuously applied. Root production on the impermeable surface is high and the large surface area traps and accumulates matter. Plant top-growth provides nutrient uptake, shade for protection against algal growth and water removal in the form of transpiration, while the large mass of self-generating root systems and accumulated material serve as living filters. Jewell et al. (1983) have hypothesised the following mechanisms, taking place in three plant sections:

- Roughing or preliminary treatment by plant species with large root systems capable of surviving and growing in a grossly polluted condition. Large sludge accumulations, anaerobic conditions and trace metal precipitation and entrapment characterise this mechanism and a large portion of wastewater BOD and suspended solids would thereby be removed.
- Nutrient conversion and recovery due to high biomass production.
- Wastewater polishing during nutrient-limited plant production, depending on the required effluent quality.

A three year pilot-scale study by Jewell et al. (1983) proved this to be a viable alternative for sewage treatment. Reed canary grass was used as the main test species and resulted in the production of better than secondary effluent quality at an application rate of 10 cm/d of settled domestic sewage and synthetic wastewater. The highest loading rates achieved were equivalent to treating the sewage generated by a population of 10,000 on an area of 2 ha. Plants other than reed canary grass were also tested and those that flourished best in the NFT system were: cattails, bulrush, strawflowers, Japanese millet, roses, Napier grass, marigolds, wheat and phragmites.

Soil-aquifer Treatment (SAT)

Where soil and groundwater conditions are favourable for artificial recharge of groundwater through infiltration basins, a high degree of upgrading can be achieved by allowing partially-treated sewage effluent to infiltrate into the soil and move down to the groundwater. The unsaturated or "vadose" zone then acts as a natural filter and can remove essentially all suspended solids, biodegradable materials, bacteria, viruses, and other microorganisms. Significant reductions in nitrogen, phosphorus, and heavy metals concentrations can also be achieved.

After the sewage, treated in passage through the vadose zone, has reached the groundwater it is usually allowed to flow some distance through the aquifer before it is collected. This additional movement through the aquifer can produce further purification (removal of microorganisms, precipitation of phosphates, adsorption of synthetic organics, etc.) of the sewage. Since the soil and aquifer are used as natural treatment. Soil-aquifer treatment is, essentially, a low-technology, advanced wastewater treatment system. It also has an aesthetic advantage over conventionally treated sewage in that water recovered from an SAT system is not only clear and odour-free but it comes from a well, drain, or via natural drainage to a stream or low area, rather than from a sewer or sewage treatment plant. Thus, the water has lost its connotation of sewage and the public see it water more as coming out of the ground (groundwater) than as sewage effluent. This could be an important factor in the public acceptance of sewage reuse schemes.

SAT System Layouts

The simplest being where the sewage effluent is applied to infiltration basins on high ground from where it moves down to the groundwater and eventually drains naturally through an aquifer into a lower area. This lower area can be a natural depression or seepage area, a stream or lake, or a surface drain. Instead of discharging wastewater directly into streams or lakes, it is applied to infiltration basins at a higher elevation so that it receives soil-aquifer treatment before entering the stream or lake.

Where the groundwater is too deep to collect the renovated sewage water by gravity, pumped wells must be used and there are two basic layouts. In one, the infiltration basins are arranged in two parallel strips and the wells are located on the line midway between the two strips. In the other, the infiltration basins are located close together in a cluster and the wells are on a circle around this cluster. Systems 11C and 11D can be used both for seasonal underground storage of sewage water, allowing the groundwater mound to rise during periods of low irrigation water demand (winter), and for pumping the groundwater mound down in periods of high irrigation water demands (summer).

While SAT systems give considerable water quality improvement to the sewage effluent, the quality of the resulting renovated water is not often as good as that of the native groundwater. Thus, SAT systems should normally be designed and managed to prevent encroachment of sewage water into the aquifer outside the portion of the aquifer used for soil-aquifer treatment. This could be achieved by ensuring that all the renovated water is intercepted by the surface or subsurface drain, which would result from excavating or installing the drain deeply enough to make sure that groundwater on the other side of the drain also moves towards the drain.

This requires groundwater-level monitoring in a few observation wells installed at the outer edges of the infiltration strips. Sewage water should travel sufficient distance through the soil and aquifer, and residence times in the SAT system should be long enough, to produce renovated water of the desired quality. While 100 m underground travel and one month underground retention time have been suggested as rule-of-thumb values, the actually required values depend on the quality of sewage effluent infiltrating into the ground, the soil types in the vadose zone and aquifer, the depth to groundwater, and the desired quality of the renovated water. Most of the quality improvement of sewage effluent moving through an SAT system occurs in the top 1m of soil. However, longer travel is desirable because it gives more complete removal of microorganisms and "polishing" treatment.

Soil Requirements

Infiltration basins for SAT systems should be located in soils that are permeable enough to give high infiltration rates. This requirement is important where sewage flows are relatively large, where excessive basin areas should be avoided (due to land cost) and where evaporation losses from the basins should be minimised. The soils, however, should also be fine enough to provide good filtration and quality improvement of the effluent as it passes through. Thus, the best surface soils for SAT systems are in the fine sand, loamy sand, and sandy loam range. Materials deeper in the vadose zone should be granular and preferably coarser than the surface soils. Soil profiles consisting of coarse-textured material on top and finer-textured material deeper down should be avoided because of the danger that fine suspended material in the sewage will move through the coarse upper material and accumulate on the deeper, finer material. This could cause clogging of the soil profile at some depth, where removal of the clogging material would be very difficult.

Vadose zones should not contain clay layers or other soils that could restrict the downward movement of water and form perched groundwater mounds. Aquifers should be sufficiently deep and transmissive to prevent excessive rises of the groundwater table (mounding) due to infiltration. Soil and aquifer materials should be granular. Fractured-rock aquifers should be protected by a soil mantle of adequate texture and thickness (at least a few metres). Shallow soils underlain by fractured rock are not suitable for SAT systems.

Hydraulic Capacity and Evaporation

Infiltration basins in SAT systems are intermittently flooded to provide regular drying periods, for restoration of infiltration rates and for aeration of the soil. Flooding schedules typically vary from 8 hours dry-16 hours flooding to 2 weeks dry-2 weeks flooding. Therefore, SAT systems should have a number of basins so that some basins can be flooded while others are drying. Annual infiltration amounts or "hydraulic loading rates" typically vary from 15 m/year to 100 m/year, depending on soil, climate, quality of sewage effluent, and frequency of basin cleaning. Thus, assuming a sewage production of 100 l/person day, a city of 100,000 people, and a hydraulic loading rate of 50 m/year, an SAT system for the entire sewage flow would require about 7.3 ha of infiltration basins. This shows that SAT systems do not necessarily require very large land areas, provided, of course, that the soils are permeable enough and the sewage is of such a quality (low suspended solids content) so as to allow high hydraulic loading rates to be maintained.

Evaporation losses from free water surfaces in dry, warm areas typically range between 1 and 2 m/year. Since the soil of infiltration basins will be mostly wet during drying, evaporation from intermittently flooded basins will be almost the same as that under continuous flooding conditions. Assuming an SAT system with a hydraulic loading rate of 50 m/year and evaporation losses of 1.5 m/year, evaporation losses would be 3 per cent of all the sewage applied which would cause a 3 per cent increase in the concentration of dissolved salts in the sewage water.

Basin Management

Bare soil is often the best condition for the bottom of infiltration basins in SAT systems. Occasional weeds are no problem but too many weeds can hamper the soil drying process, which delays recovery of infiltration rates. Dense weeds can also aggravate mosquito and other insect problems. Low water depths (about 20 cm) may be preferable to large water depths (about 1 m) because the turnover rate of sewage applied to shallow basins is faster than for deep basins of the same infiltration rate, thus giving suspended algae less time to develop in shallow basins. Suspended algae can produce low infiltration rates because they are filtered out on the basin bottom, where they clog the soil. Also, algae, being photosynthetic, remove dissolved carbon dioxide from the water, which increases the pH of the water. At high algal concentrations, this can cause the pH to rise to 9 or 10 which, in turn, causes precipitation of calcium carbonate. This cements the soil surface and results in further soil clogging and reduction of infiltration rates. Because suspended algae and soil clogging problems are reduced, shallow basins generally yield higher hydraulic loading rates than deep basins.

During flooding, organic and other suspended solids in the sewage effluent accumulate on the bottom of the basins, producing a clogging layer which causes infiltration rates to decline. Drying of the basins causes the clogging layer to dry, crack, and form curled-up flakes; the organic material also decomposes. These processes restore the hydraulic capacity so that when the basins are flooded again, infiltration rates are close to the original, high levels. However, as flooding continues, infiltration rates decrease again until they become so low that another drying period is necessary.

Depending on how much material accumulates on the bottom of infiltration basins, periodic removal of this material is necessary. Removing the material by raking or scraping is much better than mixing it with the soil with, for example, a disk harrow. The latter practice will lead to gradual accumulation of clogging materials in the top 10 or 20 cm of the soil, eventually necessitating complete removal of this layer, which could be expensive. For clean secondary sewage effluent with suspended solids concentration of 10 to 20 mg/l, flooding and drying periods can be as long as 2 weeks each, and cleaning of basin bottoms may be necessary only once a year or once every 2 years. Primary effluent, with much higher suspended solids concentration, will require a schedule which might be 2 days flooding-8 days drying, and basin bottoms might be expected to require cleaning at the end of almost every drying period. The best schedule of flooding, drying, and cleaning of basins in a given system must be evaluated by on-site experimentation.

Pretreatment

The main constituent that must be removed from raw sewage before it is applied to an SAT system is suspended solids. Reductions in BOD and bacteria are also desirable, but less essential. In the USA, there are several hundred SAT systems and, prior to land application, the sewage typically receives conventional primary and secondary treatment because that is the treatment normally prescribed before anything can be done with the effluent. Secondary treatment removes mostly biodegradable material, as expressed by the BOD, but bacteria in the soil can also degrade organic material and reduce the BOD of the sewage to essentially zero. Thus, where pretreatment is followed by SAT, primary treatment would normally be sufficient. The primary effluent would have a higher BOD and suspended solids content than secondary effluent and this would result in somewhat lower hydraulic loading rates for the SAT system and would require more frequent basin cleaning (Rice and Bouwer 1984). However, elimination of the secondary step in conventional pretreatment of the effluent would result in very significant cost savings for the overall system.

Effects

As mentioned previously, the main constitutents that must be removed from sewage effluent before it can be used for unrestricted irrigation are pathogenic organisms. Nitrogen concentration might also have to be reduced and suspended solids and biodegradable materials should perhaps be removed to protect the irrigation system or for aesthetic reasons. If the renovated water is to be used for recreational lakes or discharged into surface water, phosphorus should also be removed to prevent algal growth in the receiving water. The following sections describe how these constituents are removed or reduced in SAT systems.

Suspended Solids

After appropriate pretreatment, the suspended solids in sewage effluent are usually relatively fine and in organic form (sewage sludge, bacteria, floes, algal cells, etc.). These solids accumulate on the soil in the infiltration basins, requiring regular drying for infiltration recovery and periodic removal from the soil by raking or scraping. For loamy sands and sandy loams, few suspended solids will penetrate into the soil and then, usually, only for a short distance (a few cm, for example). In dune sands and other coarser soils, fine and colloidal suspended solids (including algal cells) can penetrate much greater distances. Except for medium and coarse uniform sands, soils are very effective filters, and suspended solids will be essentially completely removed from the sewage effluent after about 1m of percolation through the vadose zone. Additional details regarding suspended solids removal and clogging are given in Bouwer (1985) and Bouwer and Chaney (1974).

Organic Compounds

Most organic compounds of human, animal or plant origin in sewage effluent are rapidly decomposed in the soil. Under aerobic conditions (intermittent flooding), breakdown is generally faster and more complete (to carbon dioxide, minerals and water) than under anaerobic conditions. The latter prevail in the soil profile during continuous or long-term flooding. Stable, non-toxic organic compounds such as humic and fulvic acids can be formed as products of reactions between proteins and carbohydrates (cellulose or lignin). The BOD_5 of sewage varies from several hundred to about 1000 mg/l for raw sewage, and from about 10 to 20 mg/l for good quality secondary effluent. SAT systems can handle high BOD-loadings, probably hundreds of kg/ha day (Bouwer and Chaney 1974), and BOD levels are generally reduced to essentially zero after a few metres (often less) of percolation through soil. However, the final product water from SAT systems still contains some organic carbon, usually a few mg/l. This is probably mostly due to humic and fulvic acids but also to synthetic organic compounds in the sewage effluent that do not break down in the underground environment.

Halogenated hydrocarbons tend to be more resistant to biodegradation than non-halogenated hydrocarbons (Bouwer et al. 1984; Bouwer and Rice 1984). Synthetic organic compounds in the renovated water from SAT systems are generally present at very low concentrations, usually at the ppb (micrograms/l) level, and are not considered a problem when the water is used for irrigation. If it were to be used for drinking, however, additional treatment of the water by, for example, carbon filtration and reverse osmosis, would be necessary to remove the organic compounds. Additional details regarding BOD removal in SAT systems are given in Bouwer (1985) and Bouwer and Chaney (1974).

Bacteria and Viruses

Pathogenic organisms in sewage effluent include salmonella, shigella, mycobacterium, and *vibrio comma*. Specific tests for these bacteria are not routinely carried out but, instead, the numbers of faecal coliform bacteria are normally determined. *Escherichia coli* are indicator organisms that are widely used to detect faecal contamination of water and the assumption is that if faecal coliform bacteria are present in a sample, then human pathogenic bacteria could also exist. It is also inferred that if faecal coliform bacteria are no longer present, pathogenic bacteria are also absent.

Viruses in sewage effluent include entero- and adeno-viruses. Hepatitis viruses are of special concern. Viruses in renovated water from SAT systems are tested for by passing large volumes (1000 to 2000 l) through positively-charged filters to trap the viruses. Subsequently the viruses are determined in the laboratory as plaque-forming units (PFU's), which usually represent clusters of viruses. Specific viruses are tested for serologically. Other pathogens in sewage effluent include protozoa and helminth parasites, which are discussed elsewhere.

Soil is an effective filter to remove microorganisms from sewage effluent (except, of course, coarse soils such as sands and gravels, or fractured rock). Bacteria are physically strained from the water, whereas the much smaller viruses are usually adsorbed. This adsorption is favoured by a low pH, a high salt concentration in the sewage, and high relative concentrations of calcium and magnesium over monovalent cations such as sodium and potassium.

Human bacteria and viruses immobilised in the soil do not reproduce, and eventually die. Most bacteria and viruses die in a few weeks to a few months, but much longer survival times have also been reported. Many studies indicate essentially complete faecal coliform removal after percolation of 1 to a few metres through the soil. However, much longer distances of underground travel of microorganisms have also been reported. Usually, these long distances are associated with macropores, as may be found in gravelly or other coarse materials, structured or cracked clay soils, fractured rock, cavernous limestones, etc.

The best protection against breakthrough of pathogenic microorganisms in the renovated sewage water from SAT systems is to reduce bacterial levels in the sewage effluent before infiltration, to avoid coarse textured materials in the SAT systems, and to allow long underground travel distances and retention times. Additional information on this subject is provided in Bouwer (1985), Bouwer and Chaney (1974) and Gerba and Goyal (1985).

Nitrogen

Nitrogen levels in sewage can range from 20 to more than 100 mg/l, depending on in-house water use and diet of the local people and on the treatment of the sewage effluent prior to SAT. Nitrogen is primarily present as organic, ammonium, and nitrate nitrogen. The relative amounts of these nitrogen forms depend on the form of treatment prior to SAT. For secondary effluent, much of the nitrogen will often be in the ammonium form but some processes are designed to achieve nitrification and the effluent will then contain primarily nitrate-nitrogen. Raw sewage has considerable organic nitrogen.

The desirable form and concentration of nitrogen in the renovated sewage water from an SAT system depends on the nitrogen and water requirements of the crops to be irrigated, the need for preventing nitrate pollution of groundwater in the irrigated area due to excess nitrogen application to the crops, and on other possible uses of the water (including fish ponds, for which low concentrations of ammonium are required).

Control of the form and concentration of the nitrogen in renovated water from an SAT system is possible by properly selecting hydraulic loading rates and flooding and drying periods for the infiltration basins. For example, if the nitrogen in the sewage effluent is mostly in the ammonium form, short flooding periods and frequent drying of the infiltration basins (for example, 2 days flooding-5 days drying) will cause essentially complete nitrification of the ammonium in the soil, due to frequent aeration of the soil profile and resulting aerobic conditions. Thus, almost all the nitrogen in the renovated water from the SAT system will then be in the nitrate form and at concentrations about equal to the total nitrogen concentration in the sewage effluent applied to the basin.

Long flooding and drying periods (for example, 1 month flooding-1 month drying) would eventually lead to complete breakthrough of ammonium in the renovated water because of anaerobic conditions in the soil and absence of nitrification. If flooding and drying periods are of intermediate length (for example, 1 to 2 weeks flooding-1 to 2 weeks drying), there will be a succession of aerobic and anaerobic conditions in the upper part of the soil profile, which stimulates nitrification and denitrification. The latter is an anaerobic bacterial process that reduces nitrate to free nitrogen gas and oxides of nitrogen that return to the atmosphere. With this process, about 75 per cent of the nitrogen in sewage has been removed in an SAT system in Arizona, USA, with almost all of the remaining nitrogen in the renovated water occurring in the nitrate form.

Denitrification requires the presence of nitrate and organic carbon (an energy source for denitrifying bacteria) under anaerobic conditions. About 1 mg/l of organic carbon is required for each mg of nitrate nitrogen to be denitrified. If the nitrogen in the sewage is already mostly in the nitrate form and the water quite stabilised, organic carbon (as primary effluent, for example) may have to be added to the sewage effluent to achieve sufficient denitrification when the system goes anaerobic. Local experimentation is usually required to find the optimum schedule for flooding and drying, hydraulic loading, and organic carbon addition for stimulating denitrification.

Phosphorus

Sewage effluent can contain 5 to 50 mg/l phosphorus, depending on diet and water use of the local population. During pretreatment of the sewage, and in passage through the soil of the SAT system, organic phosphorus is biologically converted to phosphate. In calcareous soils and at alkaline pH, phosphate precipitates with calcium to form calcium phosphate. In acid soils, phosphate reacts with iron and aluminium oxides in the soil to form insoluble compounds. Sometimes, phosphate is initially immobilised by adsorption to the soil and then slowly reverts to insoluble forms, allowing more adsorption of mobile phosphate, etc. In clean sands with about neutral pH, phosphate can be relatively mobile. Further information is given in Bouwer (1985) and Bouwer and Chaney (1974).

Trace Elements and Salts

Sewage effluent contains a wide spectrum of other chemicals at low concentrations. These include heavy metals, fluorine, and boron. Unless these elements were already present in large concentrations in the drinking water or added to the sewage in significant amounts by industrial discharges, their concentrations in sewage are usually below the maximum limits for irrigation water (FAO 1985). Metals are significantly retained in most soils but a high pH favours immobilisation. Fluoride can form calcium fluoride, which has a very low solubility, in the soil and is also adsorbed by various soil components, especially hydrous aluminium oxides. Boron is mobile in sands and gravels but can be adsorbed on clay. Thus, SAT systems can significantly reduce the concentrations of trace elements in sewage effluent.

Total salt concentrations in sewage effluent can be several hundred mg/l higher than in drinking water. Since SAT systems generally have sandy soils, hydraulic loading rates will be much higher than evaporation losses (for example, 50 m/yr vs 1.5 m/yr). Hence, the salt concentration in the renovated water from SAT systems will be about the same as (or slightly higher than) that of the sewage effluent. If clay or organic matter is present in the soil, there will be cation adsorption and ion exchange when the SAT system is first put into operation. However, eventually, the ionic composition of the renovated sewage water will be essentially the same as that of the sewage effluent going into the ground. SAT systems do not remove salts from sewage.

Water Pollution

Water pollution is the contamination of water bodies (e.g. lakes, rivers, oceans and groundwater). Water pollution occurs when pollutants are discharged directly or indirectly into water bodies without adequate treatment to remove harmful compounds.

Water pollution affects plants and organisms living in these bodies of water. In almost all cases the effect is damaging not only to individual species and populations, but also to the natural biological communities. Water pollution is a major global problem which requires ongoing evaluation and revision of water resource policy at all levels (international down to individual aquifers and wells). It has been suggested that it is the leading worldwide cause of deaths and diseases, and that it accounts for the deaths of more than 14,000 people daily. An estimated 700 million Indians have no access to a proper toilet, and 1,000 Indian children die of diarrheal sickness every day. Some 90 per cent of China's cities suffer from some degree of water pollution, and nearly 500 million people lack access to safe drinking water. In addition to the acute problems of water pollution in developing countries, industrialised countries continue to struggle with pollution problems as well. In the most recent national report on water quality in the United States, 45 percent of assessed stream miles, 47 percent of assessed lake acres, and 32 percent of assessed bay and estuarine square miles were classified as polluted.

Water is typically referred to as polluted when it is impaired by anthropogenic contaminants and either does not support a human use, such as drinking water, and/or undergoes a marked shift in its ability to support its constituent biotic communities, such as fish. Natural phenomena such as volcanoes, algae blooms, storms, and earthquakes also cause major changes in water quality and the ecological status of water.

Categories

Surface water and groundwater have often been studied and managed as separate resources, although they are interrelated. Surface water seeps through the soil and becomes groundwater. Conversely, groundwater can also feed surface water sources. Sources of surface water pollution are generally grouped into two categories based on their origin.

Point Sources

Point source water pollution refers to contaminants that enter a waterway from a single, identifiable source, such as a pipe or ditch. Examples of sources in this category include discharges from a sewage treatment plant, a factory, or a city storm drain. The U.S. Clean Water Act (CWA) defines point source for regulatory enforcement purposes. The CWA definition of point source was amended in 1987 to include municipal storm sewer systems, as well as industrial stormwater, such as from construction sites.

Non-point Sources

Non–point source pollution refers to diffuse contamination that does not originate from a single discrete source. NPS pollution is often the cumulative effect of small amounts of contaminants gathered from a large area. A common example is the leaching out of nitrogen compounds from fertilized agricultural lands. Nutrient runoff in stormwater from "sheet flow" over an agricultural field or a forest are also cited as examples of NPS pollution.

Contaminated storm water washed off of parking lots, roads and highways, called urban runoff, is sometimes included under the category of NPS pollution. However, this runoff is typically channelled into storm drain systems and discharged through pipes to local surface waters, and is a point source. However where such water is not channelled and drains directly to ground it is a non-point source.

Groundwater Pollution

Interactions between groundwater and surface water are complex. Consequently, groundwater pollution, sometimes referred to as groundwater contamination, is not as easily classified as surface water pollution. By its very nature, groundwater aquifers are susceptible to contamination from sources that may not directly affect surface water bodies, and the distinction of point vs. non-point source may be irrelevant. A spill or ongoing releases of chemical or radionuclide contaminants into soil (located away from a surface water body) may not create point source or non-point source pollution, but can contaminate the aquifer below, defined as a toxin plume. The movement of the plume, called a plume front, may be analysed through a hydrological transport model or groundwater model. Analysis of groundwater contamination may focus on the soil characteristics and site geology, hydrogeology, hydrology, and the nature of the contaminants.

Causes

The specific contaminants leading to pollution in water include a wide spectrum of chemicals, pathogens, and physical or sensory changes such as elevated temperature and discoloration. While many of the chemicals and substances that are regulated may be naturally occurring (calcium, sodium, iron, manganese, etc.) the concentration is often the key in determining what is a natural component of water, and what is a contaminant. High concentrations of naturally-occurring substances can have negative impacts on aquatic flora and fauna.

Oxygen-depleting substances may be natural materials, such as plant matter (e.g. leaves and grass) as well as man-made chemicals. Other natural and anthropogenic substances may cause turbidity (cloudiness) which blocks light and disrupts plant growth, and clogs the gills of some fish species.

Many of the chemical substances are toxic. Pathogens can produce water-borne diseases in either human or animal hosts. Alteration of water's physical chemistry includes acidity (change in pH), electrical conductivity, temperature, and eutrophication. Eutrophication is an increase in the concentration of chemical nutrients in an ecosystem to an extent that increases in the primary productivity of the ecosystem. Depending on the degree of eutrophication, subsequent negative environmental effects such as anoxia (oxygen depletion) and severe reductions in water quality may occur, affecting fish and other animal populations.

Pathogens

Coliform bacteria are a commonly used bacterial indicator of water pollution, although not an actual cause of disease. Other microorganisms sometimes found in surface waters which have caused human health problems include:

- *Burkholderia pseudomallei*
- *Cryptosporidium parvum*
- *Giardia lamblia*
- *Salmonella*
- *Novovirus* and other viruses
- Parasitic worms (helminths).

High levels of pathogens may result from inadequately treated sewage discharges. This can be caused by a sewage plant designed with less than secondary treatment (more typical in less-developed countries). In developed countries, older cities with aging infrastructure may have leaky sewage collection systems (pipes, pumps, valves), which can cause sanitary sewer overflows. Some cities also have combined sewers, which may discharge untreated sewage during rain storms. Pathogen discharges may also be caused by poorly managed livestock operations.

Chemical and other Contaminants

Organic water pollutants include:

- Detergents
- Disinfection by-products found in chemically disinfected drinking water, such as chloroform
- Food processing waste, which can include oxygen-demanding substances, fats and grease
- Insecticides and herbicides, a huge range of organohalides and other chemical compounds
- Petroleum hydrocarbons, including fuels (gasoline, diesel fuel, jet fuels, and fuel oil) and lubricants (motor oil), and fuel combustion by-products, from stormwater runoff
- Tree and bush debris from logging operations

- Volatile organic compounds (VOCs), such as industrial solvents, from improper storage.
- Chlorinated solvents, which are dense non-aqueous phase liquids (DNAPLs), may fall to the bottom of reservoirs, since they don't mix well with water and are denser.
  — Polychlorinated biphenyl (PCBs)
  — Trichloroethylene
- Perchlorate
- Various chemical compounds found in personal hygiene and cosmetic products.

Inorganic water pollutants include:

- Acidity caused by industrial discharges (especially sulphur dioxide from power plants)
- Ammonia from food processing waste
- Chemical waste as industrial by-products
- Fertilizers containing nutrients—nitrates and phosphates—which are found in stormwater runoff from agriculture, as well as commercial and residential use
- Heavy metals from motor vehicles (via urban stormwater runoff) and acid mine drainage
- Silt (sediment) in runoff from construction sites, logging, slash and burn practices or land clearing sites

Macroscopic pollution—large visible items polluting the water—may be termed "floatables" in an urban stormwater context, or marine debris when found on the open seas, and can include such items as:

- Trash or garbage (e.g. paper, plastic, or food waste) discarded by people on the ground, along with accidental or intentional dumping of rubbish, that are washed by rainfall into storm drains and eventually discharged into surface waters
- Nurdles, small ubiquitous water-borne plastic pellets
- Shipwrecks, large derelict ships

Thermal Pollution

Thermal pollution is the rise or fall in the temperature of a natural body of water caused by human influence. Thermal pollution, unlike chemical pollution, results in a change in the physical properties of water. A common cause of thermal pollution is the use of water as a coolant by power plants and industrial manufacturers. Elevated water temperatures decreases oxygen levels (which can kill fish) and affects ecosystem composition, such as invasion by new thermophilic species. Urban runoff may also elevate temperature in surface waters. Thermal pollution can also be caused by the release of very cold water from the base of reservoirs into warmer rivers.

Transport and Chemical Reactions of Water Pollutants

Most water pollutants are eventually carried by rivers into the oceans. In some areas of the world the influence can be traced hundred miles from the mouth by studies using hydrology transport models. Advanced computer models such as SWMM or the DSSAM Model have been used in many locations worldwide to examine the fate of pollutants in aquatic systems. Indicator filter feeding species such as copepods have also been used to study pollutant fates in the New York Bight, for example.

The highest toxin loads are not directly at the mouth of the Hudson River, but 100 kilometres south, since several days are required for incorporation into planktonic tissue. The Hudson discharge flows south along the coast due to coriolis force. Further south then are areas of oxygen depletion, caused by chemicals using up oxygen and by algae blooms, caused by excess nutrients from algal cell death and decomposition. Fish and shellfish kills have been reported, because toxins climb the food chain after small fish consume copepods, then large fish eat smaller fish, etc. Each successive step up the food chain causes a stepwise concentration of pollutants such as heavy metals (e.g. mercury) and persistent organic pollutants such as DDT. This is known as biomagnification, which is occasionally used interchangeably with bioaccumulation.

Large gyres (vortexes) in the oceans trap floating plastic debris. The North Pacific Gyre for example has collected the so-called "Great Pacific Garbage Patch" that is now estimated at 100 times the size of Texas. Many of these long-lasting pieces wind up in the stomachs of marine birds and animals. This results in obstruction of digestive pathways which leads to reduced appetite or even starvation.

Many chemicals undergo reactive decay or chemically change especially over long periods of time in groundwater reservoirs. A noteworthy class of such chemicals is the chlorinated hydrocarbons such as trichloroethylene (used in industrial metal degreasing and electronics manufacturing) and tetrachloroethylene used in the dry cleaning industry (note latest advances in liquid carbon dioxide in dry cleaning that avoids all use of chemicals). Both of these chemicals, which are carcinogens themselves, undergo partial decomposition reactions, leading to new hazardous chemicals (including dichloroethylene and vinyl chloride). Groundwater pollution is much more difficult to abate than surface pollution because groundwater can move great distances through unseen aquifers. Non-porous aquifers such as clays partially purify water of bacteria by simple filtration (adsorption and absorption), dilution, and, in some cases, chemical reactions and biological activity: however, in some cases, the pollutants merely transform to soil contaminants. Groundwater that moves through cracks and caverns is not filtered and can be transported as easily as surface water. In fact, this can be aggravated by the human tendency to use natural sinkholes as dumps in areas of Karst topography.

There are a variety of secondary effects stemming not from the original pollutant, but a derivative condition. An example is silt-bearing surface runoff, which can inhibit the penetration of sunlight through the water column, hampering photosynthesis in aquatic plants.

Measurement

Water pollution may be analysed through several broad categories of methods: physical, chemical and biological. Most involve collection of samples, followed by specialised analytical tests. Some methods may be conducted *in situ,* without sampling, such as temperature. Government agencies and research organisations have published standardised, validated analytical test methods to facilitate the comparability of results from disparate testing events.

Sampling

Sampling of water for physical or chemical testing can be done by several methods, depending on the accuracy needed and the characteristics of the contaminant. Many contamination events are sharply restricted in time, most commonly in association with rain events. For this reason "grab" samples are often inadequate for fully quantifying contaminant levels. Scientists gathering this type of data often employ auto-sampler devices that pump increments of water at either time or discharge intervals. Sampling for biological testing involves collection of plants and/or animals from the surface water body. Depending on the type of assessment, the organisms may be identified for biosurveys (population counts) and returned to the water body, or they may be dissected for bioassays to determine toxicity.

Physical Testing

Common physical tests of water include temperature, solids concentrations (e.g., total suspended solids (TSS)) and turbidity.

Chemical Testing

Water samples may be examined using the principles of analytical chemistry. Many published test methods are available for both organic and inorganic compounds. Frequently used methods include pH, biochemical oxygen demand (BOD), chemical oxygen demand (COD), nutrients (nitrate and phosphorus compounds), metals (including copper, zinc, cadmium, lead and mercury), oil and grease, total petroleum hydrocarbons (TPH), and pesticides.

Biological Testing

Biological testing involves the use of plant, animal, and/or microbial indicators to monitor the health of an aquatic ecosystem.

Domestic Sewage

Domestic sewage is 99.9 percent pure water, while the other 0.1 percent are pollutants. Although found in low concentrations, these pollutants pose risk on a large scale. In urban areas, domestic sewage is typically treated by centralised sewage treatment plants. In the U.S., most of these plants are operated by local government agencies, frequently referred to as publicly owned treatment works (POTW). Municipal treatment plants are designed to control conventional pollutants: BOD and suspended solids. Well-designed and operated systems (i.e., secondary treatment or better) can remove 90 percent or more of these pollutants. Some plants have additional sub-systems to treat nutrients and pathogens. Most municipal plants are not designed to treat toxic pollutants found in industrial wastewater.

Cities with sanitary sewer overflows or combined sewer overflows employ one or more engineering approaches to reduce discharges of untreated sewage, including:

- utilising a green infrastructure approach to improve stormwater management capacity throughout the system, and reduce the hydraulic overloading of the treatment plant
- repair and replacement of leaking and malfunctioning equipment
- increasing overall hydraulic capacity of the sewage collection system (often a very expensive option).

A household or business not served by a municipal treatment plant may have an individual septic tank, which treats the wastewater on site and discharges into the soil. Alternatively, domestic wastewater may be sent to a nearby privately owned treatment system (e.g. in a rural community).

Industrial Wastewater

Some industrial facilities generate ordinary domestic sewage that can be treated by municipal facilities. Industries that generate wastewater with high concentrations of conventional pollutants (e.g. oil and grease), toxic pollutants (e.g. heavy metals, volatile organic compounds) or other non-conventional pollutants such as ammonia, need specialised treatment systems. Some of these facilities can install a pre-treatment system to remove the toxic components, and then send the partially-treated wastewater to the municipal system. Industries generating large volumes of wastewater typically operate their own complete on-site treatment systems.

Some industries have been successful at redesigning their manufacturing processes to reduce or eliminate pollutants, through a process called pollution prevention.

Heated water generated by power plants or manufacturing plants may be controlled with:

- cooling ponds, man-made bodies of water designed for cooling by evaporation, convection, and radiation
- cooling towers, which transfer waste heat to the atmosphere through evaporation and/or heat transfer
- cogeneration, a process where waste heat is recycled for domestic and/or industrial heating purposes.

Agricultural Wastewater

Nonpoint Source Controls

Sediment (loose soil) washed off fields is the largest source of agricultural pollution in the United States. Farmers may utilise erosion controls to reduce runoff flows and retain soil on their fields. Common techniques include contour plowing, crop mulching, crop rotation, planting perennial crops and installing riparian buffers.

Nutrients (nitrogen and phosphorus) are typically applied to farmland as commercial fertilizer; animal manure; or spraying of municipal or industrial wastewater (effluent) or sludge. Nutrients may also enter runoff from crop residues, irrigation water, wildlife, and atmospheric deposition. Farmers can develop and implement nutrient management plans to reduce excess application of nutrients. To minimise pesticide impacts, farmers may use Integrated Pest Management (IPM) techniques (which can include biological pest control) to maintain control over pests, reduce reliance on chemical pesticides, and protect water quality.

Point Source Wastewater Treatment

Farms with large livestock and poultry operations, such as factory farms, are called *concentrated animal feeding operations* or *confined animal feeding operations* in the U.S. and are being subject to increasing government regulation. Animal slurries are usually treated by containment in lagoons before disposal by spray or trickle application to grassland. Constructed wetlands are sometimes used to facilitate treatment of animal wastes, as are anaerobic lagoons. Some animal slurries are treated by mixing with straw and composted at high temperature to produce a bacteriologically sterile and friable manure for soil improvement.

Construction Site Stormwater

Sediment from construction sites is managed by installation of:

- erosion controls, such as mulching and hydroseeding, and
- sediment controls, such as sediment basins and silt fences.

Discharge of toxic chemicals such as motor fuels and concrete washout is prevented by use of:

- spill prevention and control plans, and
- specially designed containers (e.g. for concrete washout) and structures such as overflow controls and diversion berms.

Urban Runoff (Stormwater)

Effective control of urban runoff involves reducing the velocity and flow of stormwater, as well as reducing pollutant discharges. Local governments use a variety of stormwater management techniques to reduce the effects of urban runoff. These techniques, called best management practices (BMPs) in the U.S., may focus on water quantity control, while others focus on improving water quality, and some perform both functions.

Pollution prevention practices include low impact development techniques, installation of green roofs and improved chemical handling (e.g. management of motor fuels & oil, fertilizers and pesticides). Runoff mitigation systems include infiltration basins, bioretention systems, constructed wetlands, retention basins and similar devices.

Thermal pollution from runoff can be controlled by stormwater management facilities that absorb the runoff or direct it into groundwater, such as bioretention systems and infiltration basins. Retention basins tend to be less effective at reducing temperature, as the water may be heated by the sun before being discharged to a receiving stream.

Urban Runoff

Urban runoff is surface runoff of rainwater created by urbanisation. This runoff is a major source of water pollution in many parts of the United States and other urban communities worldwide.

Overview

Impervious surfaces (roads, parking lots and sidewalks) are constructed during land development. During rain storms and other precipitation events, these surfaces (built from materials such as asphalt and concrete), along with rooftops, carry polluted stormwater to storm drains, instead of allowing the water to percolate through soil. This causes lowering of the water table (because groundwater recharge is lessened) and flooding since the amount of water that remains on the surface is greater. Most municipal storm sewer systems discharge stormwater, untreated, to streams, rivers and bays.

Pollutants in Urban Runoff

Water running off these impervious surfaces tends to pick up gasoline, motor oil, heavy metals, trash and other pollutants from roadways and parking lots, as well as fertilizers and pesticides from lawns. Roads and parking lots are major sources of polycyclic aromatic hydrocarbons (PAHs), which are created as combustion by-products of gasoline and other fossil fuels, as well as of the heavy metals nickel, copper, zinc, cadmium, and lead. Roof runoff contributes high levels of synthetic organic compounds and zinc (from galvanised gutters). Fertilizer use on residential lawns, parks and golf courses is a significant source of nitrates and phosphorus in urban runoff.

As stormwater is channelled into storm drains and surface waters, the natural sediment load discharged to receiving waters decreases, but the water flow and velocity increases. In fact, the impervious cover in a typical city creates five times the runoff of a typical woodland of the same size.

Effects of Urban Runoff

A 2008 report by the United States National Research Council (left) identified urban runoff as a leading source of water quality problems.

Further declines in water quality remain likely if the land-use changes that typify more diffuse sources of pollution are not addressed... These include land-disturbing agricultural, silvicultural, urban, industrial, and construction activities from which hard-to-monitor pollutants emerge during wet-weather events. Pollution from these landscapes has been almost universally acknowledged as the most pressing challenge to the restoration of waterbodies and aquatic ecosystems nationwide.

—National Research Council, *Urban Stormwater Management in the United States*

The runoff also increases temperatures in streams, harming fish and other organisms. (A sudden burst of runoff from a rainstorm can cause a fish-killing shock of hot water.) Also, road salt used to melt snow on sidewalks and roadways can contaminate streams and groundwater aquifers.

One of the most pronounced effects of urban runoff is on watercourses that historically contained little or no water during dry weather periods (often called *ephemeral streams*). When an area around such a stream is urbanised, the resultant runoff creates an unnatural year-round streamflow that hurts the vegetation, wildlife and stream bed of the waterway. Containing little or no sediment relative to the historic ratio of sediment to water, urban runoff rushes down the stream channel, ruining natural features such as meanders and sandbars, and creates severe erosion—increasing sediment loads at the mouth while severely incising the stream bed upstream. As an example, on many Southern California beaches at the mouth of a waterway, urban runoff carries trash, pollutants, excessive silt, and other wastes, and can pose moderate to severe health hazards.

Because of fertilizer and organic waste that urban runoff often carries, eutrophication often occurs in waterways affected by this type of runoff. After heavy rains, organic matter in the waterway is relatively high compared with natural levels, spurring growth of algae blooms that soon use up most of the oxygen. Once the naturally occurring oxygen in the water is depleted, the algae blooms die, and in their decomposing process, cause further eutrophication. Algae blooms mostly occur in areas with still water, such as stream pools and the pools behind dams, weirs, and some drop structures. Eutrophication usually comes with deadly consequences for fish and other aquatic organisms.

Excessive stream bank erosion may cause flooding and property damage. For many years governments have often responded to urban stream erosion problems by modifying the streams through construction of hardened embankments and similar control structures using concrete and masonry materials. Use of these hard materials destroys habitat for fish and other animals. Such a project may stabilise the immediate area where flood damage occurred, but often it simply shifts the problem to an upstream or downstream segment of the stream.

Prevention and Mitigation of Urban Runoff

Effective control of urban runoff involves reducing the velocity and flow of stormwater, as well as reducing pollutant discharges. A variety of stormwater management practices and systems may be used to reduce the effects of urban runoff. Some of these techniques, called best management practices (BMPs) in the U.S., focus on water quantity control, while others focus on improving water quality, and some perform both functions.

Pollution prevention practices include low impact development techniques, installation of green roofs and improved chemical handling (e.g. management of motor fuels & oil, fertilizers and pesticides). Runoff mitigation systems include infiltration basins, bioretention systems, constructed wetlands, retention basins and similar devices.

Generation

Surface runoff can be generated either by rainfall or by the melting of snow, or glaciers. Snow and glacier melt occur only in areas cold enough for these to form permanently. Typically snowmelt will peak in the spring and glacier melt in the summer, leading to pronounced flow maxima in rivers affected by them. The determining factor of the rate of melting of snow or glaciers is both air temperature and the duration of sunlight. In high mountain regions, streams frequently rise on sunny days and fall on cloudy ones for this reason.

In areas where there is no snow, runoff will come from rainfall. However, not all rainfall will produce runoff because storage from soils can absorb light showers. On the extremely ancient soils of Australia and Southern Africa, proteoid roots with their extremely dense networks of root hairs can absorb so much rainwater as to prevent runoff even when substantial amounts of rain fall. In these regions, even on less infertile cracking clay soils, high amounts of rainfall and potential evaporation are needed to generate any surface runoff, leading to specialised adaptations to extremely variable (usually ephemeral) streams.

Infiltration Excess Overland Flow

This occurs when the rate of rainfall on a surface exceeds the rate at which water can infiltrate the ground, and any depression storage has already been filled. This is called infiltration excess overland flow, Hortonian overland flow (after Robert E. Horton), or unsaturated overland flow. This more commonly occurs in arid and semi-arid regions, where rainfall intensities are high and the soil infiltration capacity is reduced because of surface sealing, or in paved areas. This occurs largely in city areas where pavements prevent water infiltration.

Subsurface Return Flow

After water infiltrates the soil on an up-slope portion of a hill, the water may flow laterally through the soil, and exfiltrate (flow out of the soil) closer to a channel. This is called subsurface return flow or throughflow.

As it flows, the amount of runoff may be reduced in a number of possible ways: a small portion of it may evapotranspire; water may become temporarily stored in microtopographic depressions; and a portion of it may become run-on, which is the infiltration of runoff as it flows overland. Any remaining surface water eventually flows into a receiving water body such as a river, lake, estuary or ocean.

Mitigation and Treatment

Mitigation of adverse impacts of runoff can take several forms:

- Land use development controls aimed at minimising impervious surfaces in urban areas
- Erosion controls for farms and construction sites
- Flood control programmes
- Chemical use and handling controls in agriculture, landscape maintenance, industrial use, etc.

Land use controls. Many world regulatory agencies have encouraged research on methods of minimising total surface runoff by avoiding unnecessary hardscape. Many municipalities have produced guidelines and codes (zoning and related ordinances) for land developers that encourage minimum width sidewalks, use of pavers set in earth for driveways and walkways and other design techniques to allow maximum water infiltration in urban settings. An example land use control programme can be viewed at seen in the city of Santa Monica, California.

Erosion controls have appeared since medieval times when farmers realised the importance of contour farming to protect soil resources. Beginning in the 1950s these agricultural methods became increasingly more sophisticated. In the 1960s some state and local governments began to focus their efforts on mitigation of construction runoff by requiring builders to implement erosion and sediment controls (ESCs). This included such techniques as: use of straw bales and barriers to slow runoff on slopes, installation of silt fences, programming construction for months that have less rainfall and minimising extent and duration of exposed graded areas. Montgomery County, Maryland implemented the first local government sediment control programme in 1965, and this was followed by a statewide programme in Maryland in 1970.

Flood control programmes as early as the first half of the twentieth century became quantitative in predicting peak flows of riverine systems. Progressively strategies have been developed to minimise peak flows and also to reduce channel velocities. Some of the techniques commonly applied are: provision of holding ponds (also called detention basins) to buffer riverine peak flows, use of energy dissipators in channels to reduce stream velocity and land use controls to minimise runoff.

Chemical use and handling. Following enactment of the U.S. Resource Conservation and Recovery Act (RCRA) in 1976, and later the Water Quality Act of 1987, states and cities have become more vigilant in controlling the containment and storage of toxic chemicals, thus preventing releases and leakage. Methods commonly applied are: requirements for double containment of underground storage tanks, registration of hazardous materials usage, reduction in numbers of allowed pesticides and more stringent regulation of fertilizers and herbicides in landscape maintenance. In many industrial cases, pretreatment of wastes is required, to minimise escape of pollutants into sanitary or stormwater sewers.

The U.S. Clean Water Act (CWA) requires that local governments in urbanised areas (as defined by the Census Bureau) obtain stormwater discharge permits for their drainage systems. Essentially this means that the locality must operate a stormwater management programme for all surface runoff that enters the municipal separate storm sewer system ("MS4"). EPA and state regulations and related publications outline six basic components that each local programme must contain:

- Public education (informing individuals, households, businesses about ways to avoid stormwater pollution)
- Public involvement (support public participation in implementation of local programmes)
- Illicit discharge detection & elimination (removing sanitary sewer or other non-stormwater connections to the MS4)
- Construction site runoff controls (i.e. erosion & sediment controls)
- Post-construction (i.e. permanent) stormwater management controls
- Pollution prevention and "good housekeeping" measures (e.g. system maintenance).

Other property owners which operate storm drain systems similar to municipalities, such as state highway systems, universities, military bases and prisons, are also subject to the MS4 permit.

Peak Water

The term Peak Water has been put forward as a concept to help understand growing constraints on the availability, quality, and use of freshwater resources. The clearest definitions of the term were laid out in a 2010 peer-reviewed article in the Proceedings of the National Academy of Sciences by Peter Gleick and Meena Palaniappan. They define peak renewable, peak non-renewable, and peak ecological water. There is a vast amount of water on the planet but sustainably managed water is becoming scarce. An updated assessment was published in August 2011 in the Stockholm International Water Institute's journal. Much of the world's water in underground aquifers and in lakes behaves like a finite resource by being depleted. The phrase *peak water* sparks debates similar to those about peak oil. The concept has sparked such high interest that The New York Times chose the phrase "peak water" as one of their 33 "Words of the Year" for 2010.

There is concern that the state of peak water is being approached in many areas around the world. Some areas are suffering from peak renewable water, where entire renewable flows are being consumed for human use, peak non-renewable water, where groundwater aquifers are being overpumped (or contaminated) faster than nature recharges them (this example is most like the peak oil debate), and peak ecological water, where ecological and environmental constraints are overwhelming the economic benefits provided by water use. If present trends continue, 1.8 billion people will be living with absolute water scarcity by 2025, and two thirds of the world population could be subject to water stress. Ultimately, peak water is not about running out of fresh water, but about reaching physical, economic, and environmental limits on meeting human demands for water and the subsequent decline of water availability and use.

Peak Water Comparison with Peak Oil

The Hubbert curve has gained a high degree of popularity in the scientific community for predicting the depletion of various natural resources. M. King Hubbert created this measurement device in 1956 for a variety of finite resources such as coal, oil, natural gas and uranium. Hubbert's curve was not applied to resources such as water originally, since water is a renewable resource.

Some forms of water, however, such as fossil groundwater, exhibit similar characteristics to oil, and overpumping (faster than the rate of natural recharge of groundwater) often results in a Hubbert-type peak. A modified Hubbert curve applies to any resource that can be harvested faster than it can be replaced. Like peak oil, peak water is inevitable given the rate of extraction of certain water systems. A current argument is that growing populations and demands for water will inevitably lead to non-renewable use of water resources.

Water Supply

Fresh water is a renewable resource, yet the world's supply of clean, fresh water is under increasing demand for human activities. The world has an estimated 1.34 billion cubic kilometres of water, but 96.5 percent of it is salty. Nearly 70 per cent of that fresh water is frozen in the icecaps of Antarctica and Greenland. Most of the remainder is present as soil moisture or lies in deep underground aquifers as groundwater not accessible to human use. Less than 1 per cent of the world's fresh water or 0.007 per cent of all water on earth is accessible for direct human use. This is the water found in lakes, rivers, reservoirs and those underground sources that are shallow enough to be tapped at an affordable cost. Only this minuscule amount is regularly renewed by rain and snowfall, and is therefore available on a sustainable basis.

The amount of available freshwater supply in some regions is decreasing because of (i) climate change, which has caused receding glaciers, reduced stream and river flow, and shrinking lakes; (ii) contamination of water by human and industrial wastes; and (iii) overuse of non-renewable groundwater aquifers. Many aquifers have been over-pumped and are not recharging quickly. Although the total fresh water supply is not used up, much has become polluted, salted, unsuitable or otherwise unavailable for drinking, industry and agriculture.

Table: *Freshwater sources (top 15 countries)*

Country	*Total Freshwater Supply (km^3/yr)*
Brazil	8233
Russia	4498
Canada	3300
Colombia	3132
USA	3069
Indonesia	2838
China	2830
Peru	1913
India	1908
Cong	1283
Venezuela	1233
Bangladesh	1211
Burma	1046
Chile	922
Vietnam	891

Water Demand

Water demand already exceeds supply in many parts of the world, and as the world population continues to rise, many more areas are expected to experience this imbalance in the near future.

Agriculture represents 70 per cent of freshwater use worldwide.

Agriculture, industrialisation and urbanisation all serve to increase water consumption.

Fresh Water Withdrawal by Country

The highest total annual consumption of water comes from India, China and the United States, countries with large populations and extensive agricultural irrigation demand for food production. See the following table:

Table: *Freshwater withdrawal by country and sector (top 20 countries)*

Country	*Total Freshwater Withdrawal (km^3/yr)*	*Per Capita Withdrawal (m^3/p/yr)*	*Domestic Use (%)*	*Industrial Use (%)*	*Agricultural Use (%)*
India	645.84	585	8	5	86
China	549.76	415	7	26	68
United States	477	1,600	13	46	41
Pakistan	169.39	1,072	2	2	96

Japan	88.43	690	20	18	62
Indonesia	82.78	372	8	1	91
Thailand	82.75	1,288	2	2	95
Bangladesh	79.4	560	3	1	96
Mexic	78.22	731	17	5	77
Russia	76.68	535	19	63	18
Iran	72.88	1,048	7	2	91
Vietnam	71.39	847	8	24	68
Egypt	68.3	923	8	6	86
Brazil	59.3	318	20	18	62
Uzbekistan	58.34	2,194	5	2	93
Canada	44.72	1,386	20	69	12
Iraq	42.7	1,482	3	5	92
Italy	41.98	723	18	37	45
Turkey	39.78	544	15	11	74
Germany	38.01	460	12	68	20

India

India has 20 percent of the Earth's population, but only four per cent of its water. Water tables are dropping fast in some of India's main agricultural areas. The mighty Indus and Ganges rivers are tapped so heavily that, except in rare wet years, they no longer reach the sea.

India has the largest water withdrawal out of all the countries in the world. Eighty-six per cent of that water goes to support agriculture. That heavy use is dictated in large part by what people eat. People in India consume a lot of rice. Rice farmers in India typically get less than half the yield per unit area and they use 10 times more water than their counterparts in China do.

Economic development in some ways makes matters worse. As people's living standards rise, they tend to eat more meat. But it takes tremendous quantities of water to raise animals for food. Growing a tonne of grain requires 1,000 tonnes of water. To produce a tonne of beef takes 15,000 tonnes. To make a single hamburger requires around 4940 litres (1,300 gallons) of water A glass of orange juice needs 850 litres (225 gallons) of fresh water to produce.

China

China, as the most populous country in the world, has the second largest water withdrawal out of all the countries in the world. Sixty-eight per cent of that water goes to support agriculture and its growing industrial base is consuming twenty-six percent. China is facing a water crisis where water resources are over-allocated, inefficiently used, and grossly polluted by human and industrial wastes. One third of China's population lacks access to safe drinking water. Rivers and lakes are dead and dying, groundwater aquifers are over-pumped, uncounted species of aquatic life have been driven to extinction, and direct adverse impacts on both human and ecosystem health are widespread and growing.

In western China's Qinghai province, through which the Yellow River's main stream flows, more than 2,000 lakes have disappeared over the last 20 years. There were once 4,077 lakes. Global climate change is responsible for the reduction in flow of the (Huang He) Yellow River over the past several decades. The source of the Yellow River is the Qinghai-Xizang Tibetan plateau where the glaciers are receding sharply.

In Hebei Province, which surrounds Beijing, the situation is much worse. Hebei is one of China's major wheat and corn growing provinces. The water tables have been falling fast throughout Hebei. The region has lost 969 of its 1,052 lakes. About 500,000 people are affected by a shortage of drinking water due to continuing droughts. Hydro-power generation is also impacted. Two large cities - Beijing and Tianjin depend on Hebei Province to supply their water from the Yangtze River. Beijing gets its water via the newly constructed South-North Water Transfer Project. The river originates in a glacier on the eastern part of the Tibetan plateau.

United States

The United States has about 5 per cent (1/20) of the world's population, yet the US uses almost as much water as India (~1/5 of world) or China (1/5 of world) because substantial amounts of water are used to grow food exported to the rest of the world. The agricultural sector in the US consumes more water than the industrial sector, though substantial quantities of water are withdrawn (but not consumed) for power plant cooling systems. There are 36 states in the U.S. in some form of water stress, from serious to severe.

The Ogallala Aquifer in the southern high plains (Texas and New Mexico) is being mined at a rate that far exceeds replenishment—a classic example of peak non-renewable water. Portions of the aquifer will not naturally recharge due to layers of clay between the surface and the water-bearing formation, and because rainfall rates simply do not match rates of extraction for irrigation. The term fossil water is sometimes used to describe water in aquifers that was stored over centuries to millennia. Use of this water is not sustainable when the recharge rate is slower than the rate of groundwater extraction.

In California, massive amounts of groundwater are also being withdrawn from Central Valley groundwater aquifers — unreported, unmonitored, and unregulated. California's Central Valley is home to one sixth of all U.S. irrigated land, and the state leads the nation in agricultural production and exports. The inability to sustain groundwater withdrawals over time may lead to adverse impacts on the region's agricultural productivity. The Central Arizona Project (CAP) is a 336-mile (541 km) long canal that diverts 489 billion US gallons (1.85×10^2 m^3) a year from the Colorado River to irrigate more than 300,000 acres (1,200 km^2) of farmland. The CAP project also provides drinking water for Phoenix and Tucson. It has been estimated that Lake Mead, which dams the Colorado, has a 50-50 chance of running dry by 2021. The Ipswich River near Boston now runs dry in some years due to heavy pumping of groundwater for irrigation. Maryland, Virginia and the District have been fighting over the Potomac River. In drought years like 1999 or 2003, and on hot summer days the region consumes up to 85 percent of the river's flow.

Per Capita Withdrawal of Water

Turkmenistan, Kazakhstan and Uzbekistan use the most water per capita. See the table below:

Table: *Freshwater withdrawal by Country and Sector (top 15 countries, per capita)*

Country	*Total Freshwater Withdrawal (km^3/yr)*	*Per Capita Withdrawal (m^3/p/yr)*	*Domestic Use (%)*	*Industrial Use (%)*	*Agricultural Use (%)*
Turkmenistan	24.65	5,104	2	1	98
Kazakhstan	35	2,360	2	17	82
Uzbekistan	58.34	2,194	5	2	93
Guyana	1.64	2,187	2	1	98
Hungary	21.03	2,082	9	59	32
Azerbaijan	17.25	2,051	5	28	68
Kyrgyzstan	10.08	1,916	3	3	94
Tajikistan	11.96	1,837	4	5	92
USA	477	1,600	13	46	41
Suriname	0.67	1,489	4	3	93
Iraq	42.7	1,482	3	5	92
Canada	44.72	1,386	20	69	12
Thailand	82.75	1,288	2	2	95
Ecuador	16.98	1,283	12	5	82
Australia	24.06	1,193	15	10	75

Turkmenistan

Turkmenistan, gets most of its water from the Amu Darya River. The Qaraqum Canal is a canal system that takes water from the Amu Darya River and distributes the water out over the desert for irrigation of its orchard crops and cotton. Turkmenistan uses the most water per capita in the world because only 55 per cent of the water delivered to the fields actually reaches the crops.

Kazakhstan and Uzbekistan

The two rivers feeding the Aral Sea were dammed up and the water was diverted to irrigate the desert so that cotton could be produced. As a result, the Aral Sea's water has become much saltier and the sea's water level has decreased by over 60 per cent. Drinking water is now contaminated with pesticides and other agricultural chemicals and contains bacteria and viruses. The climate has become more extreme in the area surrounding it.

Water Shortfall by Country

Saudi Arabia, Libya, Yemen and United Arab Emirates have hit peaks in water production and are depleting their water supply. See the table below:

Table: *Freshwater Shortfall by Country (Top 15 Countries)*

Region & Country	*Total Freshwater Withdrawal (km^3/yr)*	*Total Freshwater Supply (km^3/yr)*	*Total Freshwater Shortfall (km^3/yr)*
Saudi Arabia	17.32	2.4	-14.9
Libya	4.27	0.6	-3.7
Yemen	6.63	4.1	-2.5
United Arab Emirates	2.3	0.2	-2.2
Kuwait	0.44	0.02	-0.4
Oman	1.36	1.0	-0.4
Israel	2.05	1.7	-0.4
Qatar	0.29	0.1	-0.2
Bahrain	0.3	0.1	-0.2
Jordan	1.01	0.9	-0.1
Barbados	0.09	0.1	0.0
Maldives	0.003	0.03	0.0
Antigua and Barbuda	0.005	0.1	0.0
Malta	0.02	0.07	0.1
Cyprus	0.21	0.4	0.2

Saudi Arabia

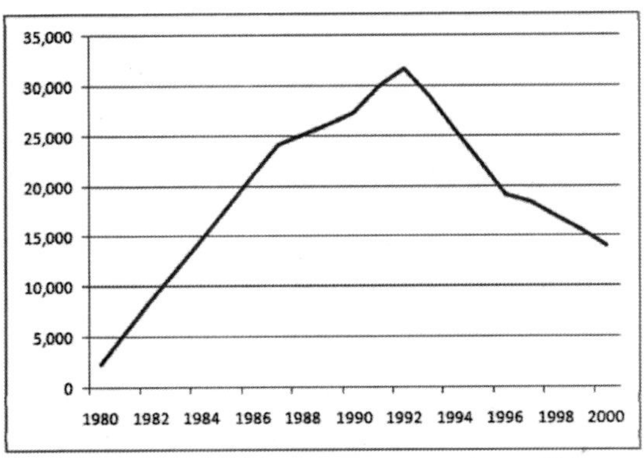

Figure: *Water supply in Saudi Arabia, 1980–2000, in millions of cubic meters*

According to Walid A. Abderrahman (2001), "Water Demand Management in Saudi Arabia", Saudi Arabia reached peak water in the early 1990s, at more than 30 billion cubic meters per year, and declined afterwards. The peak had arrived at about midpoint, as expected for a Hubbert curve.

Today, the water production is about half the peak rate. Saudi Arabian food production has been based on "fossil water" - water from ancient aquifers that is being recharged very slowly if at all. Fossil water is non-renewable, just as oil is, and it is unavoidable that it has to run out someday.

Saudi Arabia has abandoned its self-sufficient food production and is now importing all of its food. Saudi Arabia has built desalination plants to provide about half the country's fresh water. The remainder comes from groundwater (40%), surface water (9%) and reclaimed wastewater (1%).

Libya

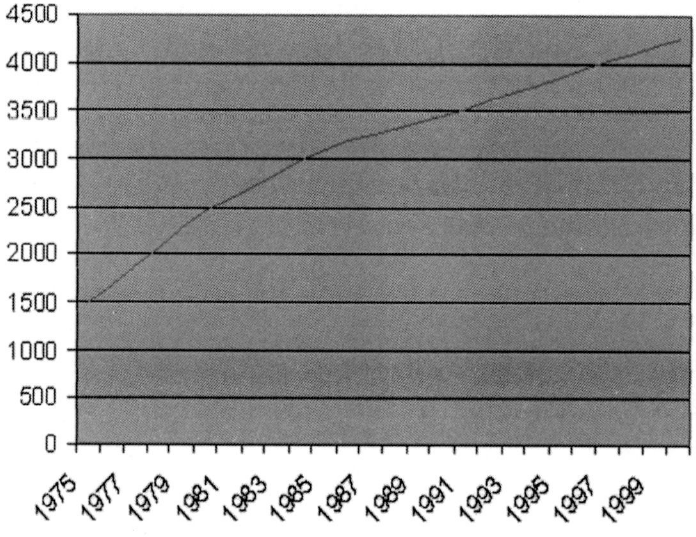

Figure: *Water supply in Libya, 1975–2000, in millions of cubic meters*

Libya is working on a network of water pipelines to import water, called the Great Manmade River. It carries water from wells tapping fossil water in the Sahara desert to the cities of Tripoli, Benghazi, Sirte and others. Their water also comes from desalination plants.

Yemen

Peak water has occurred in Yemen. Sustainability is no longer attainable in Yemen, according to the government's five-year water plan for 2005-2009. The aquifer that supplies Sana'a, the capital of Yemen, will be depleted by 2009. In its search for water in the basin, the Yemeni government has drilled test wells that are 2 kilometres (1.2 mi) deep, depths normally associated with the oil industry, but it has failed to find water. Yemen must soon choose between relocating the city and building a pipeline to coastal desalination plants.

United Arab Emirates

United Arab Emirates has a rapidly growing economy and very little water to support it. UAE requires more water than is naturally available. They have reached peak water. To solve this, UAE has a desalination plant near Ruwais and ships its water via pipeline to Abu Dhabi.

Consequences

Famine

Water shortage may cause famine in Pakistan. Pakistan has approximately 35 million acres (140,000 km^2) of arable land irrigated by canals and tube wells, mostly using water from the Indus River. Dams were constructed at Chashma, Mangla, and Tarbela to feed the irrigation system.

Since the completion of the Tarbela Dam in 1976 no new capacity has been added despite astronomical growth in population. The gross capacity of the three dams has decreased because of sedimentation, a continual process. Per-capita surface-water availability for irrigation was 5,260 cubic meters per year in 1951. This has been reduced to a mere 1,100 cubic meters per year in 2006. The water shortage will cause a wheat deficit of 12 million tonnes per year by 2012-13.

Health Problems

The quality of drinking water is vital for human health. Peak water constraints result in people not having access to safe water for basic personal hygiene. "Infectious water-borne diseases such as diarrhea, typhoid, and cholera are responsible for 80 percent of illnesses and deaths in the developing world, many of them children. One child dies every eight seconds from a water-borne disease; 15 million children a year."

Vital aquifers everywhere are becoming contaminated with toxins. Once an aquifer is contaminated, it is not likely that it can ever recover. Contaminants are more likely to cause chronic health effects. Water can be contaminated from pathogens such as bacteria, viruses, and parasites.

Also, toxic organic chemicals can be a source of water contamination. Inorganic contaminants include toxic metals like arsenic, barium, chromium, lead, mercury, and silver. Nitrates are another source of inorganic contamination. Finally, leaching radioactive elements into the water supply can contaminate it.

Human Conflicts over Water

Some large or small conflicts of the future may be fought over some aspects of water, including availability, quality, and control. Water has also been used as a tool in conflicts or as a target during conflicts that start for other reasons. Water shortages may well result in water conflicts over this precious resource.

In West Africa today and in many other places, Nepal, Bangladesh, India (such as the Ganges-Brahmaputra), and Peru, major changes in the rivers, generate a significant risk of violent conflict in coming years. Water management and control could well play a part in future resource wars between contending parties desiring scarce resources.

Issues Defy Easy Solutions

No matter how freshwater is used, whether for agriculture, industry, or municipalities, there is great potential for better conservation and management. Water is used inefficiently nearly everywhere. Until actual scarcity hits, many take access to freshwater for granted. Water resources can be properly managed and water can be conserved. And, sometimes "backstop" water sources are viable.

Water Conservation

Water conservation refers to reducing the use of water. There are several things that can be done to conserve water. Most irrigation systems waste water. Typically, only between 35 per cent and 50 per cent of water withdrawn for irrigated agriculture ever reaches the crops. Most soaks into unlined canals, leaks out of pipes, or evaporates before reaching (or after being applied to) the fields. Swales and cisterns can be used to catch excess rainwater and store it for the dry season. Water should be used more efficiently in industry. Industry should use a closed water cycle if possible. Also, industry should prevent polluting water so that it can be returned into the water cycle.

Whenever possible gray waste water should be used to irrigate trees or lawns. Water drawn from aquifers should be recharged by treating the wastewater and replacing it back into the aquifer. Finally, water can be conserved by not allowing fresh water to be used to irrigate luxuries such as golf courses. Luxury goods should not be produced in areas where fresh water has been depleted. For example, 1500 litres of water is used on average for the manufacturing of a single computer and monitor.

Water Management

Sustainable water management involves the scientific planning, developing, distribution and optimum utilising of water resources under defined water polices and regulations. Examples of policies that improve water management include the use of technology for efficiency monitoring and use of water, innovative water prices and markets, irrigation efficiency techniques, and much more. Experience shows that higher water prices lead to improvements in the efficiency of use - a classical argument in economics, pricing, and markets. Many examples support this approach. In 2008, Clark County, Nevada, raised its water rates to encourage conservation. Economists propose to encourage conservation by adopting a system of progressive pricing whereby the price per unit of water used would start out very small, and then rise substantially for each additional unit of water used. This approach - tiered rate structures - has been used for many years in many places, and is becoming more widespread. A Freakonomics column in The New York Times similarly suggested that people would respond to higher water prices by using less of it, just as they respond to higher gasoline prices by using less of it. The Christian Science Monitor has also reported on arguments that higher water prices curb waste and consumption.

Conversely, certain kinds of subsidies can lead to inefficient use of water. Water subsidies often involve contentious policy issues that are political in nature. In 2004, the Environmental Working Group criticised the U.S. federal government for selling subsidised water to corporate farms for an average price of only $17 per acre foot (326,000 gallons).

Simon wrote, "... there is complete agreement among water economists that all it takes to ensure an adequate supply for agriculture as well as for households in rich countries is that there be a rational structure of water law and market pricing. The problem is not too many people but rather defective laws and bureaucratic interventions; freeing up markets in water would eliminate just about all water problems forever... In poor water-short countries the problem with water supply - as with so many other matters - is lack of wealth to create systems to supply water efficiently enough. As these countries become richer, their water problems will become less difficult...". This theoretical argument, however, ignores strong barriers to open water markets, including the difficult of moving water from one region to another, water rights laws that prevent redistribution based solely on economic value, and grossly imperfect information on water use. And indeed, actual experience with peak water limits in some wealthy, but water-short countries and regions still suggests serious difficulties in reducing water challenges.

Climate Change

Extensive research has shown the direct links between water resources, the hydrologic cycle, and climatic change. As climate changes, there will be substantial impacts on water demands, precipitation patterns, storm frequency and intensity, snowfall and snowmelt dynamics, and more. Evidence from the IPCC Working Group II, has shown climate change is already having a direct effect on animals, plants and water resources and systems. Some findings of a 2007 report (Climate Change 2007: Climate Change Impacts, Adaptation and Vulnerability) include : 75-250 million people across Africa could face water shortages by 2020. Crop yields could increase by 20 per cent in East and Southeast Asia, but decrease by up to 30 per cent in Central and South Asia. Agriculture fed by rainfall could drop by 50 per cent in some African countries by 2020. A wide range of other impacts could affect peak water constraints.

Loss of biodiversity can be attributed largely to the appropriation of land for agroforestry and the effects of climate change. The 2008 IUCN Red List warns that long-term droughts and extreme weather puts additional stress on key habitats and, for example, lists 1,226 bird species as threatened with extinction, which is one-in-eight of all bird species.

Backstop Water Sources

The concept of a "backstop" resource is a resource that is sufficiently abundant and sustainable to replace non-renewable resources. Thus, solar and other renewable energy sources are considered "backstop" energy options for unsustainable fossil fuels. Similarly, Gleick and Palaniappan defined "backstop water sources" to be those resources that can replace unsustainable and non-renewable use of water, albeit typically at a higher cost. The classic backstop water source is desalination of seawater. If the rate of water production is not sufficient in one area, another "backstop" could be increased interbasin transfers, such as pipelines to carry freshwater from where it is abundant to an area where water is needed. Water can be imported into an area using water trucks. The most expensive and last resort measures of getting water to a community such as desalination, water transfers are called "backstop" water sources. Fog catchers are the most extreme of backstop methods.

To produce that fresh water, it can be obtained from ocean water through desalination. A January 17, 2008 article in the Wall Street Journal stated, "World-wide, 13,080 desalination plants produce more than 12 billion US gallons (45,000,000 m^3) of water a day, according to the International Desalination Association." Israel is now desalinising water at a cost of US$0.53 per cubic meter. Singapore is desalinising water for US$0.49 per cubic meter. After being desalinised at Jubail, Saudi Arabia, water is pumped 200 miles (320 km) inland though a pipeline to the capital city of Riyadh.

However, there are several factors that currently keep desalination from being the cure-all for water shortages: the high capital costs to build the desalination plant, the high cost of the water produced, the energy required to desalinate the water, the environmental issues with the disposal of the resulting brine and the high cost of transporting water. Nevertheless, some countries such as Spain are increasingly relying on desalination because of the continuing decreasing costs of the technology.

At last resort, it is possible in some particular regions to harvest water from fog using nets. The water from the nets drips into a tube. The tubes from several nets lead to a holding tank. Using this method, small communities on the edge of deserts can get water for drinking, gardening, showering and clothes washing. Critics say that fog catchers work in theory but have not succeeded as well in practice. This is due to the high expense of the nets and pipe, high maintenance and low quality water.

An alternative approach is that of the Seawater Greenhouse which consists in desalinating seawater thorugh evaporation and condensation inside a greenhouse using solar energy as the only energy supply. Succeful pilots have been conducted growing crops in desertic locations and a commercial operation is currently running in Australia.

Trophic State Index

The quantities of nitrogen, phosphorus, and other biologically useful nutrients are the primary determinants of a body of water's trophic state index (TSI). Nutrients such as nitrogen and phosphorus tend to be limiting resources in standing water bodies, so increased concentrations tend to result in increased plant growth, followed by corollary increases in subsequent trophic levels. Consequently, a body of water's trophic index may sometimes be used to make a rough estimate of its biological condition. Although the term "trophic index" is commonly applied to lakes, any surface water body may be indexed.

Carlson's Trophic State Index

Carlson's index is one of the more commonly used trophic indices, and is the trophic index used by the United States Environmental Protection Agency. The *trophic state* is defined as the total weight of biomass in a given water body at the time of measurement. Because they are of public concern, the Carlson index uses the algal biomass as an objective classifier of a lake or other water body's trophic status. According to the US EPA, the Carlson Index should only be used with lakes that have relatively few rooted plants and non-algal turbidity sources.

Index Variables

Because they tend to correlate, three independent variables can be used to calculate the Carlson Index: *chlorophyll pigments*, *total phosphorus* and *Secchi depth*. Of these three, chlorophyll will probably yield the most accurate measures, as it is the most accurate predictor of biomass. Phosphorus may be a more accurate of a water body's summer trophic status than chlorophyll if the measurements are made during the winter. Finally, the Secchi depth is probably the least accurate measure, but also the most affordable and expedient one. Consequently, citizen monitoring programmes and other volunteer or large-scale surveys will often use the Secchi depth. By translating the Secchi transparency values to a log base 2 scale, each successive doubling of biomass is represented as a whole integer index number. The Secchi depth, which measures water transparency, indicates the concentration of dissolved and particulate material in the water, which in turn can be used to derive the biomass. This relationship is expressed in the following equation:

$$\left(\frac{1}{z}\right)\left(ln\frac{I_0}{I_z}\right) = k_w + \alpha C$$

where z = the depth at which the disk disappears,

I_0 is the intensity of light striking the water's surface,

I_z is about 10 per cent of I_0 and is considered a constant,

k_w is a coefficient for the attenuation of light by water and dissolved substances,

α is treated as a constant with the units of square meters per milligram and

C is the concentration of particulate matter in units for milligrams per cubic meter.

Bibliography

Anderson, Mary P. and William W. Woessner: *Applied Groundwater Modeling: Simulation of Flow and Advective Transport.* Academic Press, San Diego, 1992.

Asaithambi, S. *: Economics of Ground Water Management in India*, Abhijeet, Delhi, 2008.

Bhakar, S.R. *: Ground Water Hydrology : Theory and Practice*, Agrotech, Delhi, 2009.

Bhave, P.R. and R. Gupta: *Analysis of Water Distribution Networks*, Narosa, Delhi, 2011.

Biswas Asit K. *: Integrated Water Resources Management in South and South-East Asia*, , Oxford University Press, Delhi, 2001.

Bourne, Peter G.: *Water and Sanitation: Economic and Sociological Perspectives* Academic Press. Orlando. 1984.

Brooks, D.: *Water: Local-Level Management*, Ottawa, International Development Research Centre, 2002.

Cullis, A. : *Rainwater Harvesting: The Collection of Rainfall and Runoff in Rural Areas,* London, U.K.: IT Publications, 1986.

Erik Nissen-Peterson : *Rainwater Catchment Systems*, UK: Intermediate Technology Publications, 1999.

Feder, Paul A.: *A Guidebook to Groundwater: Resources and Education Opportunities in the Great Lakes Region*, Great Lakes Commission, Ann Arbor, 1993.

Fetter, Charles Willard: *Applied Hydrogeology*, Merrill Publishing Company, Columbus, OH. 1988.

Geraghty, James J.: *Water Atlas of the United States*, Port Washington, NY, 1973.

Gischler, Christiaan E.: *Water Resources in the Arab Middle East and North Africa*, Cambridge, Eng, Middle East & North African Studies Press, 1979.

Gurjar, Ram Kumar : *Geography of Water Resources*, Rawat, Delhi, 2008.

Hemenway, Toby : *Gaia's Garden: A Guide to Home-Scale Permaculture*, Vermont: Chelsea Green Publishing Company, 2000.

Jat, M.L. and S.R. Bhakar: *Ground Water Hydrology : Theory and Practice*, Agrotech, Delhi, 2009.

Kanmony, J. Cyril : *Drinking Water Management : Problems and Prospects*, Mittal Pub, Delhi, 2010.

LeRoy, K. : *Sustaining Water Washington*, D.C.: Population Services International, New York, 1993.

McGarry, M.G.: *Matching Water Supply Technology to the Needs and Resources of Developing Countries,* United Nations. New York. 1987

Meenu Bhatnagar: *Groundwater Management : Sustainable Approaches*, Icfai Books, Delhi, 2012.

Morris, Mary E.: *Water Scarcity and Security concerns in the Middle East*, Abu Dhabi, Emirates Center for Strategic Studies and Research, 1998.

Narwani, G.S. : *Community Water Management*, Rawat, Delhi, 2005.

Nielsen, David M.: *Practical Handbook of Groundwater Monitoring*. Lewis Publishers, Chelsea, 1991.

Palanisami, K : *Groundwater Management and Policies*, Macmillan Publishers India, Delhi, 2008.

Premjit Sharma: *Agricultural Drainage and Water Quality*, Gene Tech Books, Delhi, 2007.

Singh, A.K.: *Environment and Water Resources Management*, Adhyayan, Delhi, 2006.

Thomas, Harold E.: *The Conservation of Groundwater: A Survey of the Present Groundwater Situation in the U.S.,* McGraw Hill, New York, NY. 1951.

Todd, David Keith: *Groundwater Hydrology*. John Wiley & Sons, New York, NY. 1959.

William, K. : *Irrigation Investment Technology, and Management Strategies for Development Studies in Water Policy Management 9*, Boulder, Colo Westview Press, 1986.

Index

M

N

O

P

R

S

T

W

❑❑❑